Names

Joore, Peter, editor. | Overdiek, Anja, editor. | Smeenk, Wina, editor. | Turnhout, Koen van, editor.

Title

Applied Design Research in Living Labs and Other Experimental Learning and Innovation Environments / edited by Peter Joore, Anja Overdiek, Wina Smeenk, Koen van Turnhout.

Description

First edition. | Boca Raton, FL: CRC Press, 2024. | Includes bibliographical references.

Funding

This publication is a result of the NADR3 project, executed by the Network Applied Design Research. This project was co-funded by Taskforce SIA, part of the Netherlands Organisation for Scientific Research (NWO). Open access of this publication was financed by NWO.

Contributors

Peter Joore, Anja Overdiek, Wina Smeenk, Koen van Turnhout, Daan Andriessen, Maria Arias, Marije Boonstra, Aranka Dijkstra, Morgan Duta, Jeroen van den Eijnde, Maaike Harbers, Donagh Horgan, Tiwánee van der Horst, Mark Jacobs, Tomasz Jaśkiewicz, Peter Joore, Elise van der Laan, Ju Laclau Massaglia, Catelijne van Middelkoop, Masi Mohammadi, Manon Mostert-van der Sar, Ryan Pescatore Frisk, Perica Savanović, Janneke Sluijs, Iskander Smit, Guido Stompff, Sybrith M. Tiekstra, Peter Troxler, Marieke Zielhuis

Design and layout

Studio RATATA.nl

Illustrations

Kalle Wolters

NETWORK
APPLIED
DESIGN
RESEARCH

NDR2401_CRC_EN20

Content

Applied Design Research in Living Labs and Other Experimental Learning and Innovation Environments

Experimental learning and innovation environments, such as living labs, field labs, and urban innovation labs, are increasingly used to connect multi-stakeholders in envisioning, creating, experimenting, learning, and trying out novel responses to diverse societal challenges. With designers facilitating lab processes and/or testing artifacts together with users, the design discipline plays an important role in these labs.

Applied Design Research in Living Labs and Other Experimental Learning and Innovation Environments combines a focus on experimental learning and innovation environments (or Living Labs) with a focus on applied design research. It offers an interdisciplinary perspective by bringing together diverse stakeholders from different disciplines; the book will adopt an interdisciplinary perspective, integrating insights from design, innovation, sociology, technology, and other relevant fields. It showcases real-world examples and case studies of successful applied design research in experimental learning and innovation environments and focuses on design dilemmas that emerge while working in experimental learning and innovation environments. The book explores the roles of various stakeholders, including the roles that may play out during the development of these labs, and goes on to discuss the balance between fixed or fluid roles of these stakeholders and the polarity between working within one specific discipline versus working with various expertise or disciplines.

Designers, government representatives, and researchers who apply a Living Lab approach to solving multi-stakeholder challenges in various fields by applying energy, mobility, health, education, or social living labs will find this book of interest.

CRC Press
Taylor & Francis Group

Edited by
Peter Joore, Anja Overdiek,
Wina Smeenk and Koen van Turnhout

Colophon

First edition published 2025
by CRC Press
2385 Executive Center Drive, Suite 320, Boca Raton, FL 33431, USA
and by CRC Press
4 Park Square, Milton Park, Abingdon, Oxon, OX14 4RN

CRC Press is an imprint of Taylor & Francis Group, LLC

Library of Congress Cataloging-in-Publication Data

Names: Joore, Peter, 1967- editor. | Overdiek, Anja, editor. | Smeenk, Wina, editor. | Turnhout, Koen van, editor.
Title: Applied design research in living labs and other experimental learning and innovation environments / edited by Peter Joore, Anja Overdiek, Wina Smeenk, Koen van Turnhout.
Description: First edition. | Boca Raton, FL : CRC Press, 2025. | Includes bibliographical references.
Identifiers: LCCN 2024020734 (print) | LCCN 2024020735 (ebook) | ISBN 9781032793191 (hardback) | ISBN 9781032792507 (paperback) | ISBN 9781003491484 (ebook)
Subjects: LCSH: Education--Experimental methods. | Instructional systems--Design. | Learning strategies. | Experiential learning.
Classification: LCC LB1026 .A66 2025 (print) | LCC LB1026 (ebook) | DDC 371.3--dc23/eng/20240722
LC record available at https://lccn.loc.gov/2024020734
LC ebook record available at https://lccn.loc.gov/2024020735

ISBN: 978-1-032-79319-1 (hbk)
ISBN: 978-1-032-79250-7 (pbk)
ISBN: 978-1-003-49148-4 (ebk)

DOI: 10.1201/9781003491484

Publisher's note: This book has been prepared from camera-ready copy provided by the authors.

Part 3: Materialisation of the Living Lab

In Conclusion

Preface

How can we increase the impact of applied design research? How can we (further) substantiate the way applied design research works (and doesn't work) precisely, and the way it can be deployed as effectively as possible to make a meaningful contribution to the complex questions facing society? Finding answers to these questions was one of the reasons the Network Applied Design Research (NADR) was established in 2016 as a national platform. What makes NADR unique is that it brings together professors, designers and researchers within vastly different knowledge and application areas, crossing sector boundaries. These experts focus on issues within healthcare, construction, the energy sector, the food and agriculture sector, retail, hospitality, and fashion, among others. In all these sectors, they employ applied design research, expecting that this unique approach can help to make a significant contribution to the development of a smart, sustainable, and inclusive society.

The unifying theme of NADR is the methodological approach to the combination of designing and researching, in every possible variant. This means that NADR is interested in Research Through Design, Research For Design, as well as Research Into Design. Through NADR's extensive network, the methodology is not only developed and applied in one or a few knowledge domains, but across many areas. By connecting knowledge and experience from different sectors, the methodological expertise in applied design research can further evolve. This leads to both strengthened methodological development, as well as to greater societal impact. NADR's ambition is to further expand this unique cross-sectoral perspective.

The target audience of NADR's expertise is the 'designing professional'. This expert may work within the creative industry (for example, at a design agency focusing on product design or service design), but also within a design or innovation department of a manufacturing company, a social organisation, or the government. Through the implementation of applied design research, the designing professional can address complex issues in collaboration with relevant stakeholders.

NADR operates on a yearly knowledge cycle, delving deeper each year into a selection of one of the methodologies within the field. Each yearly cycle concludes with a joint symposium held during the annual Dutch Design Week in Eindhoven. Subsequently, the results are documented in a publication, which you currently hold in your hands. This publication is the outcome of the 2023-2024 yearly cycle. The book comprises 13 chapters written by 28 authors and focuses on applied design research in Living Labs and other experimental learning and innovation environments.

After reading this book, you will have a good understanding of how Living Labs and similar concepts can be effectively deployed, along with their constraints and prerequisites. However, the research is by no means complete, as every study inevitably leads to new questions. Perhaps these questions will be addressed in the next yearly cycle of NADR, which focuses on the societal impact of applied design research, and how it can be deployed to enhance systemic change and contribute to societal transitions towards, for example, the circular economy, smarter healthcare, or a sustainable food system.

But one thing at a time... At this moment, as the chair of the Network Applied Design Research, I would like to extend my sincere gratitude to my fellow editors, Anja Overdiek, Wina Smeenk, and Koen van Turnhout. Together with all the authors, they have brought this book to fruition through substantive discussions and a thorough peer review process. Because one image often speaks louder than a thousand words, this NADR publication is again decorated with inspiring illustrations by Kalle Wolters, and visually designed by Studio RATATA. I am particularly proud of the outcome we have achieved, and I am confident that it will serve as inspiration for everyone wanting to find out how applied design research in Living Labs can help to accomplish a lasting positive impact on society!

Peter Joore,

Chair, Network of Applied Design Research

Koen van Turnhout, Peter Joore,
Anja Overdiek, Wina Smeenk

1. Experimentation at the Heart of Societal Change

Editorial

DOI: 10.1201/9781003491484

Introduction

Today's applied design researchers are where the action is. They have always operated as a catalyst of change: propelling and steering design and society into brighter futures (Andriessen & Van Turnhout, 2023). But now, more than ever, design researchers teamup with a broad range of stakeholders to get things done. They surround themselves with engineers, entrepreneurs, policy-makers, professionals, citizens, and take leadership in sparking ideas, making things tangible, creating meaning, and driving change. In the wake of the many societal transitions needed to combat today's challenges, everyone needs to change their ways, their thinking, and the structures they reside in. Applied design researchers can play a crucial role in making that effort worthwhile. They can be facilitators that help others to bridge boundaries of spheres of life, disciplines, sectors, and domains (Smeenk, 2022).

This book is about Living Labs and other Experimental Learning and Innovation Environments such as Field Labs and Urban Innovation Labs. These are spaces deliberately created to try new things and learn together (Overdiek & Geerts, 2021; 2023). Although a variety of terms may be used, each with their relevant differences, in this chapter we will use the term Living Labs as an overarching concept. In Chapter 2, we will further discuss how we define the concept, in relation to related concepts.

In Living Labs, municipalities may meet with entrepreneurs to improve the safety of a neighbourhood. Designers may engage with citizens based on technology probes, unleashing the creative potential of the public. Maker spaces may form the culprit of novel connections and collaborations. Music festivals may be employed both as models for society and as a safe space to explore different ways of organising it. In all of these environments, applied design researchers can play many roles, acting as initiators, developers, discoverers, change managers, networkers, interpreters, and reframers (Joore et al., 2018). In doing so, they foster safety, creativity, encouragement, and reflection. They conceptualise, materialise, test, and organise meaningful conversations among stakeholders.

But how do they go about this? What are the challenges that applied design researchers face in their efforts to promote experimentation that is valuable for obtaining new insights, imagining alternative futures, developing new knowledge, and accelerating transitions in society? And how do they navigate these challenges? What is their unique contribution and how can they improve? This book is the result of a collective reflection process on the practices of a wide range of applied design researchers united in the Network Applied Design Research (see Joore et al., 2022). It highlights their work in the Netherlands and furthers the field by sharing practices, reflections, and open questions.

Three Interconnected Perspectives

This book is structured into three parts, each focussing on a specific aspect of the Living Lab. Part 1 of this book explores Living Labs and their relationship with societal change processes, which can be considered as the macro level of our research. Living Labs are often set up in – or close to – reality, such as a neighbourhood or a company's work floor, but they also intend to provide safe havens where practical constraints are temporarily set aside, for innovative ideas to develop freely. Applied design researchers must balance creative exploration within the lab, with the practical realities of implementation in the external world. They need to transcend the temporality of the lab to connect to the larger transitions it tries to stimulate. It is necessary to translate lessons from the lab to the real world and to communicate the value of the results of the lab. How do applied design researchers navigate this delicate balance between creativity and pragmatism, transitioning new ideas from the lab into the complexities of the outside world?

Part 2 of this book is focussed on the social dynamics that take place in the Living Lab itself, which can be considered as the intermediate or meso level of our research. Applied design researchers need to take into account different perspectives, encouraging participants of the lab to collaborate across different spheres of life, disciplines, sectors, and domains. One challenge is to promote interdisciplinary teamwork while respecting each other's interests, expertise, experience, and influence (e.g., Smeenk, 2023). How can the setup of the Living Lab (the setting, the space, the participants, its activities, and ways of working) contribute to planned spontaneity? How do we get participants to share their interests, exchange ideas, and cross social, cultural and economic boundaries? And how can design researchers intervene to facilitate this process?

Part 3 of this book is focussed on the actual practices of applied design researchers within a Living Lab, for instance with regards to their engagement with tangible prototypes as boundary objects, which can be considered as the micro level of our investigation. In this part, we discuss how we can foster boundary-crossing in experimental learning environments using tangible materials. Objects such as prototypes may prompt dialogues and discussions on alternative futures, reveal values, beliefs, mental models, worldviews, and enhance stakeholders' shared understanding. They can form tangible bites for thought. However, choosing and developing the right prototype for each situation is a nuanced process, including a balance between the material, mental, and social aspects of design conversations. How do we make this work?

The book's central premise is that the impact of a Living Lab primarily depends on how applied design researchers relate these different scales or levels to each other. The lab's success depends crucially on the ability of design researchers to tie the lab to the real world, manage the social dynamics within the lab, and develop practices that positively influence those dynamics. These relations need to be seen, acknowledged, established, imagined, built, tested, and nurtured or questioned by, amongst others, applied design researchers.

The three themes of the book may be depicted as a concentric set of challenges, as presented in Figure 1. We materialise to facilitate productive social dynamics in a Living Lab to ultimately change society. One could examine this set of relationships from the inside out, starting with prototypes, and seeing how they facilitate social dynamics that lead to changes in the real world, or the other way around. We have made the – somewhat arbitrary – choice for the latter. That means that this book is organised from the outside in, starting with the relationship of Living Labs to the changes in society, followed by the relation between the setup of the lab and its internal social dynamics, and ending with the potential of materialisation in the labs to strengthen these dynamics.

But before we start diving deeper into the relationship between applied design researchers and Living Labs, we will first set the stage. For that purpose, in Chapter 2: *'Living Labs and Other Experimental Environments – Dynamics and Directions',* Anja Overdiek and Elise van der Laan present an overview of the state of the discussions regarding Living Labs. The chapter gives a historical account of Living Labs tracing how the concept emerged and evolved. They sketch the potential of Living Labs to drive societal change, made possible through the real-life characteristics and multidisciplinary approach in Living Labs. At the same time, the value of such labs still needs to be proven. Therefore, Overdiek and van der Laan set an agenda for applied design research related to Living Labs, focusing on co-design methodologies, new types of transition labs, reliability, scalability, ethics, and learning in experimental environments. The ultimate goal is to harness the potential of Living Labs for systemic change and address complex societal challenges through innovative design approaches.

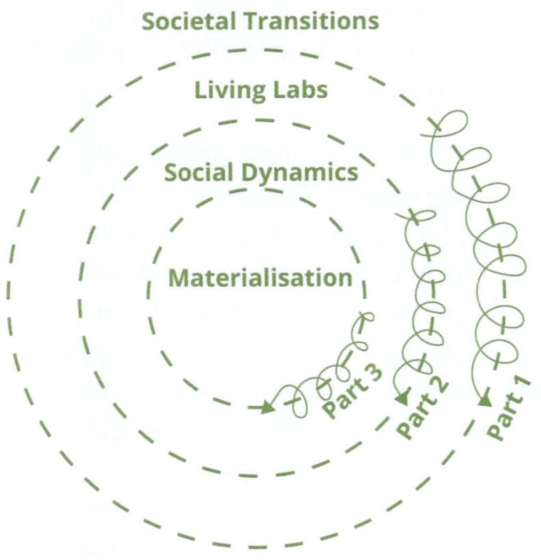

Figure 1. Three parts of the book.

Part 1: Living Labs and Societal Transitions

How do Living Labs contribute to societal change? Despite the real-life settings of many labs, they are in a way also separated from society. This is because Living Labs often are a temporary construction in a constantly evolving society, and also because they are purposely set up locally and as a safe space of trust, new rules, and tempered expectations to conduct experiments together harmlessly. How can Living Labs be set up so that they contribute to societal transitions in the best way? And what challenges do design researchers face in doing so? The chapters in this part tackle such questions.

In Chapter 3: *'The Art of Connection. Beyond the Borders of Safe Zones: a Living Lab Case Study'*, Jeroen van den Eijnde and Masi Mohammadi discuss the challenges they face in managing the expectations of diverse stakeholders involved in a Living Lab. They focus on a project called 'The Art of Connection', which explores the potential of interactive public spaces in fostering social interaction among residents, including vulnerable elderly individuals, in a City of Arnhem neighbourhood. They reflect on how constraints in the way this lab is set up may complicate the uptake of some of its results in society. For example, funding conditions, co-financing dynamics, and time-related project financing raise questions about ownership and sustainability. They describe how, despite these complexities, they try to bridge the gap between abstract project goals and the daily lives of local residents.

This question of how labs contribute to society is picked up further in Chapter 4: *'Co-designing towards Transitions? Facilitating More Than Experiments in Living Labs'*. In this chapter, Janneke Sluijs, Maria Arias, Morgan Duta, Ju Laclau Massaglia, and Anja Overdiek start from a critique on the impact of many design interventions on broader transition processes, emphasising the need for explicit connections between local experiments and larger transformations. The authors argue that designers can go beyond individual interventions by becoming 'convenors' in transition processes, leading multi-stakeholder networks through a Living Lab methodology focused on collective knowledge creation, imagination, testing, and implementation. They propose designers' suitability to co-design with networks for transformative adoption, particularly in the food, energy, and healthcare sectors and suggest an approach and a particular set of skills for this.

Going even further, Aranka Dijkstra, Peter Joore, Sybrith Tiekstra, and Marije Boonstra treat the question of how labs contribute to societal change as an empirical one. In Chapter 5: *'Exploring the Potential of Festivals as Living Labs for Systemic Innovation – Insights from the Interdisciplinary Innovation Program DORP'*, they evaluate the potential of festivals as spaces where one could experiment with systemic change. As temporary mini-societies, festivals present systemic sustainability challenges, making them ideal for experimenting with sustainable system innovations in areas like water, energy, housing, logistics, waste management, food, and behaviour. Using the Living Lab Activity Framework (LAFF), they track a number of projects and their progress across different innovation stages and system levels within the Festival Living Lab. The analysis suggests that the approach facilitates learning across various system levels, supporting the idea that festivals can serve as effective Living Labs for diverse types of innovation, including sustainable system innovation.

Part 2: Social Dynamics in Living Labs

Drilling one layer deeper, we should wonder how the lab can be organised to be the safe space needed to foster and nurture fresh ideas and novel approaches to societal challenges. How can the social dynamics between the stakeholders in a lab be fostered? What settings, spaces, participants, and ways of working contribute to the free exchange of ideas and the crossing of cultural boundaries between the lab members? How do we plan spontaneity of the lab (Van Turnhout et al., 2017)? The three chapters in this part suggest there are multiple ways forward.

In Chapter 6: *'Opening & Closing Hours – Three Cases and their Dynamics to let Learning Thrive'* , Wina Smeenk, Perica Savanović, Marieke Zielhuis, and Daan Andriessen explore collaboration dynamics in experimental learning environments. Multi-actor environments can be challenging and the authors investigate at what point a more open or a more closed approach is appropriate. They use boundary-crossing theory as a theoretical lens. The authors suggest that in many cases, an imbalance of the learning mechanisms (reflection, identification, coordination, and transformation) occurs. The dynamics of open or closed Living Lab moments form an important consideration. One should not only look at being open or closed to the participation of stakeholders, but also consider other open and/or closed aspects such as goals, approaches, design spaces, information availability, and decision-making.

Koen van Turnhout and Daan Andriessen provide a different perspective on the importance of social dynamics in labs in Chapter 7: *'Experimenting with Novel Knowledge: A Plea for Communities of Practice'*. They raise the question of how learning environments can be organised so that the individual learning of participants adds up to knowledge that transcends the realities of the lab. Building on two Communities of Practice (CoP), the Design Science Research Group and Workplace for Musal Research, they reflect on the intricacies of CoP design and management. They show how CoPs can be effective for individual learning, but how collective learning is more challenging and embedded in the epistemic cultures of the participants. They propose to integrate CoPs into normal Living Labs, for example, by forming guilds.

Next, in Chapter 8: *'The Open Lab as Boundary Object'*, Peter Troxler and Manon Mostert-van der Sar delve into the social dynamics of the makerspace. They suggest the MakerLab is set up as a third space and can be compared to a boundary object, stimulating serendipity, interdisciplinary collaboration, and innovation. The MakerLab exemplifies inclusive educational approaches, promoting interdisciplinary learning and social integration. According to Troxler and Mostert-van der Sar, open labs can serve as arenas for addressing societal challenges through applied design research and provide a space for experimentation, knowledge sharing, and cultural engagement in higher education and beyond.

Part 3: Materialisation within the Living Lab

When asked how exactly they further the collective thought in a multi-stakeholder environment, many applied design researchers will point to the value of material artefacts, either in the role of prototype or in another form. Conversations about things and objects are simply different and often more productive than conversations without such tangible references. But how does this work exactly? And what are the corresponding requirements for the materialisation itself? Authors in this section do embrace tangibility, while at the same time arguing to look beyond the traditional notion of prototypes.

First, in Chapter 9: *'Bridging Multi-Stakeholder Dialogue about AI Systems in the Lab: How Virtual Can We Go?'*, Tiwánee van der Horst, Anja Overdiek, and Maaike Harbers focus on tangible artefacts that mitigate the difficulties in developing responsible applications of emerging technologies, particularly Artificial Intelligence (AI). They introduce a designed boundary object as a tool to facilitate conversations about systemic issues related to AI's ecological and societal impacts. The study compares the effectiveness of a designed boundary object in physical and virtual contexts, revealing that the physical experience was more cohesive and effective in bridging perceptions and facilitating multi-stakeholder alignment. The authors conclude that physical experience might be crucial, specifically in the first phase of a Living Lab where multi-stakeholder dialogue must be bridged. In later phases of the lab, virtual interactions might be more viable.

Somewhat closer to traditional prototypes, but still very explorative, are technology probes (Hutchinson et al., 2003). In Chapter 10: *'Between Experiments – Leveraging Prototypes to Trigger, Articulate, and Share Informal Knowledge – Case of the Cities of Things Living Lab'*, Tomasz Jaśkiewicz and Iskander Smit argue that prototypes can be a means of explorative co-creation. It explores 'civic robots' as tangible boundary objects in connecting diverse stakeholders and fostering emergent knowledge generation. The Cities of Things Lab 010's initiative involves co-creating neighbourhood robot prototypes called *'WijkBots'*, to disrupt the commercial-driven trend in emerging robot technologies and prioritise community needs. These *'civic robots'* enhance stakeholder communication and understanding, including academia, industry, government, and citizens. This chapter emphasises the vital role of prototypes in sustaining Living Lab initiatives, aligning diverse interests, and driving sustainability within networked labs.

Guido Stompff, Mark Jacobs, and Donagh Horgan delve into the meaning of the term prototype itself. In Chapter 11: *'Ceci n'est pas un Prototype – Prototypes and Idealtypes as Representations of What Works and What Matters'*, they examine the impact of prototypes on the design thinking process. Students and coaches lacking design expertise, struggle with the production and refinement of prototypes as a working method. Surprisingly, stakeholders vary significantly in their view of what a 'prototype' represents.

For some, it signifies the final design outcome, focusing on 'what works'; for others, it embodies the ideal, emphasising 'what should be'. The authors propose clarifying the term '*prototype*' for representing what can be and introducing '*ideal type*' for representations corresponding to the ideal, fostering separate discussions on 'what matters' and 'what works', especially relevant in social innovation and social design where diverse stakeholder perspectives and values exist.

In Chapter 12: *'Concerning Apples & Oysters'*, Catelijne van Middelkoop and Ryan Pescatore Frisk put the material aspect of Living Labs in a much broader, historical, and educational context. Working from the history of art education in Europe, they show how the practices at SintLucas connect educational, cultural, and technological thinking through materiality. They discuss 'Transition Atelier The Last Makers', a Living Lab concept that facilitates collaboration between humans and non-humans, fostering innovation and learning in a real-world setting. It explores boundary materials to generate ideas and alternative futures, emphasising long-term problem-solving and sustainable decision-making through (un)making processes. Collaboration challenges require ongoing reflection for effective knowledge development and education.

Conclusion

All of the following chapters will discuss the unique contribution of applied design researchers to Living Labs and other Experimental Learning and Innovation Environments. By exploring this subject from different perspectives, ranging from the macro level focussing on the relationship of the Living Lab and the overarching societal transitions, to the meso level related to the social dynamics within the Living Lab itself, to the microlevel of specific material practices, the reader will discover how these levels are interconnected, and how applied design researchers may support these vital and indispensable connections.

Bibliography

- Andriessen, D., & van Turnhout, K. (2023). Stromingen in Ontwerpgericht Onderzoek. In K. van Turnhout, D. Andriessen, & P. Cremers (Eds.), *Handboek Ontwerpgericht Wetenschappelijk Onderzoek*. Boom Uitgeverij.
- Joore, P., van den Eijnde, J., van't Veer, J. T., van Beurden, K., Hadzhivalcheva, I., Brinks, G., & Mohammadi, M. (2018). *In Joore, P., Stompff, G., & van den Eijnde, J. (Eds.). (2022). Applied Design Research: A Mosaic of 22 Examples, Experiences, and Interpretations Focussing on Bridging the Gap Between Practice and Academics. CRC Press.*
- Joore, P., Stompff, G., & van den Eijnde, J. (Eds.). (2022). *Applied Design Research: A Mosaic of 22 Examples, Experiences, and Interpretations Focussing on Bridging the Gap Between Practice and Academics*. CRC Press.
- Hutchinson, H., Mackay, W., Westerlund, B., Bederson, B. B., Druin, A., Plaisant, C., ... Eiderbäck, B. (2003, April). Technology probes: inspiring design for and with families. In *Proceedings of the SIGCHI conference on Human factors in computing systems* (pp. 17-24).
- Overdiek, A., & Geerts, H. (Eds.). (2021). *Innovating with Labs: That's how You Do It!; Insights from Future-Proof Retail*. The Hague University of Applied Sciences.
- Overdiek, A., & Geerts, H. (Eds.). (2023). *Innovating with Labs 2.0. Creating space for sustainability transitions*. The Hague University of Applied Sciences.
- Smeenk, W. (2022). A Systemic Co-Design Iceberg: A systemic perspective in the ever-evolving practice of empathic co-design. *RSD Conference*.
- Smeenk, W. (2023). The empathic Co-Design Canvas: A tool for supporting multi-stakeholder co-design processes. *International Journal of Design, 17*(2), 81-98.
- Van Turnhout, K., Annema, J. H., van de Goor, J., Marjolein, J., & Bakker, R. (2017). Planning Spontaneity: a case study about method configuration. In *Analysing Design Thinking: Studies of Cross-Cultural Co-Creation* (pp. 389-403). CRC Press.

Anja Overdiek,
Elise van der Laan

2. Living Labs and Other Experimental Environments

Dynamics and Directions

DOI: 10.1201/9781003491484

II.
I.
III.

Introduction

Municipalities collaborate with citizens to find locally adapted ways to deal with the energy transition. Industry clusters work together with regional governments and knowledge institutions to develop and test new sustainable materials. Universities co-create with students and industry to find a more practice-oriented and experimental manner of learning. Nowadays everyone is developing environments to innovate and collaborate with diverse actors to face challenges that have outgrown the potential of just one group of actors tackling them. Most of these experimental environments and even the projects running in them use the term 'Living Lab' to describe themselves. Where does this popular approach come from and how did it develop in practice? How do different sorts of real-life labs relate to each other? And in what directions could particularly designers and applied design research help to develop experimental environments further? These are some questions the following chapter will unpack to provide an overview of the dynamic landscape of Living Labs and other experimental environments. We will do this based on a review of the international literature on Living Labs and on our own experience as practice researchers in this field in the Netherlands.

The Origins of Living Labs

Currently, we can speak of an abundance of experimentation environments. From the end of the 20th century onwards, companies started taking innovation outside their confined research and development (R&D) departments in search of 'new solutions to problems that go beyond the boundaries of individual organisations or institutions' (Paskaleva & Cooper, 2021, p.2). This development can be traced back to the rise of two research methods: 'open innovation' in the early 2000s (Chesbrough, 2003) and user-driven or user-centred innovation that came up as early as the 1990s (Ballon & Schuurman, 2015; Dell'Era & Landoni, 2014). Unlike traditional closed innovation, where companies rely primarily on their internal R&D departments, open innovation encourages collaboration with external partners, including customers, suppliers, universities, research institutions, and even competitors. Collaborating with a diverse group of stakeholders brings different and rich perspectives to problem-solving. This can lead to more comprehensive and effective solutions to complex challenges. User-Centred Design (UCD; Norman & Draper, 1985) or user-driven innovation (Von Hippel, 2005) are approaches to designing products, services or IT systems that focus on their end-users. These approaches require understanding the users' perspectives, involving them in the design process, and iterating on designs based on their feedback. Nowadays, the user-centred approach has broadened to understanding human needs from the very beginning of the design process. The common term for this is Human-Centred Design (HCD) which has even solidified in the field of Human-ComputerInteraction as an ISO norm (https://www.iso.org/standard/77520.html).

The 'Living Lab' has developed as a popular approach to collaborate with diverse actors and involve users in innovation. A Living Lab is thought to bridge open innovation with user-centred innovation through experimentation (Leminen et al., 2012; Ballon & Schuurman, 2015). The concept finds its roots in different contexts that revolve around co-creation, the main ones being the Scandinavian movement of cooperative and participative design in the 1960s and 1970s (Sanoff, 1990), European social experiments with IT in the 1980s, and digital city initiatives in the 1990s that introduced the first public virtual spaces in which technology users were considered innovative agents (Ballon & Schuurman, 2015; Mastelic et al., 2015). Building on these traditions, the term Living Lab emerged in the early 2000s for data-gathering methods studying users in real-life settings, either in a real-life like (digital) laboratory or in users' everyday habitat. In Europe, research through Living Labs grew into a full-fledged movement through the impetus of European policy measures and funding schemes, marked by the launch of the European Network of Living Labs (ENoLL) in 2006.

Increasingly, Living Labs have also been employed in the public sector as instruments in developing knowledge and solutions relating to societal 'wicked' or complex problems (Paskaleva & Cooper, 2021; Voytenko et al., 2016), using a triple or quadruple helix perspective. Here, governments work together with knowledge institutions, the private sector, and citizens to find new approaches and develop, test, and sometimes implement 'solutions'.

A Family of Open Laboratories

Although a unified definition for Living Labs is lacking, the general description offered by EnOLL in 2006 is much referred to: 'Living Labs are user-centred open innovation eco-systems based on [a] systematic user co-creation approach, integrating research and innovation processes in real life communities and settings'[1]. In other words, Living Labs are multi-stakeholder networks for collaboration on open-innovation that involves (certain groups of) employees or citizens in real-life environments to co-create, test, and validate new technologies, services, products, and societal innovations.

In the past two decades, Living Labs have become a widespread phenomenon, with diverse interpretations and research across different disciplines (Greve et al., 2021). Inspired by the research and innovation theories described in the section before, the Living Lab has mostly been a practice phenomenon. Some refer to the physical environment in which experiments take place as a Living Lab, while others use the term to describe the overarching approach of a research project, of which (short-term) real-life experiments are a part. In practice as in literature, there is a lack of consensus on what environments and projects are considered a Living Lab, how they should be organised, and what methods are used. Next to Living Lab, many terms circulate for open, real-life laboratories, such as: Fieldlabs (FiL), Social Innovation Labs (SIL), Testlabs (TL), Fablabs (FaL), Festival Labs (FeL), and Urban Living Labs (ULL), to name just the most frequently mentioned lab types in publications. To make it more puzzling, some methodologies which include experimentation environments, like Transition Arena's (TA) inspired by Transition Management (Loorbach & Rotmans, 2010; Notermans et al., 2022) do not call themselves Living Labs. On the other hand, some learning environments which are experienced as experimental but do not co-create, test, and validate innovations such as Communities of Practice (CoP) and Hybrid Learning Communities (HLC) sometimes call themselves labs.

---

1 In literature up until at least 2017 (see for example Steen & van Bueren, 2017), authors commonly refer to the definition offered by European Network of Living Labs (EnOLL) on their website since 2006. This has in recent years been updated and now reads, 'Living Labs (LLs) are open innovation ecosystems in real-life environments using iterative feedback processes throughout a lifecycle approach of an innovation to create sustainable impact. They focus on co-creation, rapid prototyping & testing and scaling-up innovations & businesses, providing (different types of) joint-value to the involved stakeholders. In this context, Living Labs operate as intermediaries/orchestrators among citizens, research organisations, companies and government agencies/levels', retrieved on February 28, 2024 via https://enoll.org/about-us/what-are-living-labs/.

Experimentation environments in the Netherlands have been classified according to their distance to daily life and to their level of co-creation between different stakeholders (Maas et al., 2017). They could also be distinguished by their permanence. Some like the Festival Lab (FeLs) are organised temporarily, whereas many Fieldlabs (FiL) like those in the energy or digital industries are more permanent institutions, governed by public-private partnerships.

Finally an important distinction between experiments over the years has been their focus on techno/managerial goals versus a focus on social/civic goals (Sengers et al., 2019; Alavi et al., 2017; Westley et al., 2017). This distinction coincides with the difference between experimental environments in practice, for example the difference between Fieldlabs (FiL) and Social Innovation Labs (SILs). Both lab environments work on the innovation of socio-technical systems, but the first focus more on technical innovation, whereas the second aim at social innovation. This different focus also reflects in the main (or initiating) actors in these labs: industry players and entrepreneurs in the first, and citizens and (often local) governments in the second.

However, in order to make an impact, it seems very important that different labs 'learn from each other, build on each other and develop a shared knowledge base together' (Broek et al., 2020, p. 7). Rather than pin down each lab type with its own characteristics, we thus propose to look at the Living Lab via the lens of the family metaphor. This implies that there is a core similarity, but also lots of differences between the different above-mentioned labs. Core elements constituting the ideal type of a Living Lab can arguably be summarised as (Ballon & Schuurman, 2015):

1. user involvement
2. real-life setting for experimentation
3. multi-stakeholder participation
4. multi-method approach, drawing on disciplines including ethnography and sociology
5. co-creation between different stakeholders

As pointed out earlier, Living Labs are both physical or virtual environments (places) and projects geared to specific experiments which temporarily build these environments. That's why the core elements we suggest above do not draw on the place-project distinction, but rather on the real-life and multi-stakeholder approach, which is characterised by the core elements above. Based on the 'family resemblance' with this approach, we can draw a circle around the Living Lab (LL) visualising which experimental environments are close and which might be further away from the family core. We are including the names of experimental environments which we see most often in recent literature and practice in the Netherlands (see Figure 1), but, of course, others could be added.

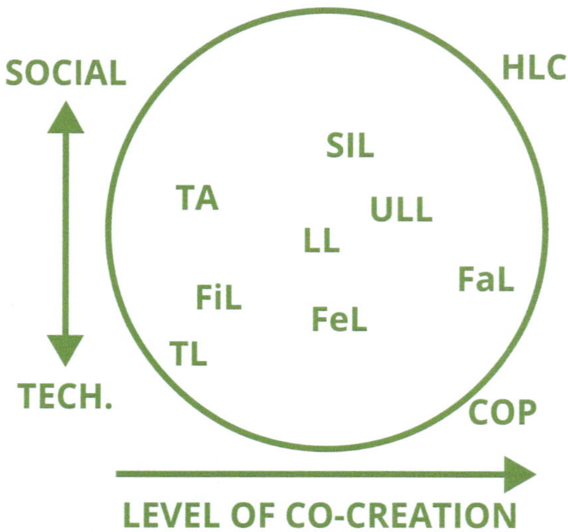

Figure 1. Family circle of open experimental environments.

Figure 1 shows the ideal type Living Lab (LL) positioned in the family core, symbolising experimental environments which are characterised by all five core elements. The lesser of this core elements a lab embraces, the further it is situated towards the edge of this 'family' circle, with labs only using one core element placed on the border. We also suggest using the continuum between less or more inclusive co-creation on an imagined x-axis and the continuum between technological and social innovation goals on an imagined y-axis. Figure 1 suggests how current lab types could be situated but foremost invites labs to position themselves in this family circle and thus transparently reflect on their core elements.

Impact of Living Labs

Substantial claims on the impact of Living Labs have been made, including their ability to tap into tacit knowledge (Leminen & Westerlund, 2012; Dell'Era & Landoni, 2014; Pierson & Lievens, 2005), to enable innovative collaboration and ideas (Leminen et al., 2012) and the effective involvement of multiple stakeholders (Paskaleva & Cooper, 2021; Greve et al., 2021). Yet, as some studies point out, there is scant evidence that Living Labs realise these benefits. Do they actually deliver a higher level of co-creation than more closed experimentation environments (Steen & van Bueren, 2017; Greve et al., 2017)? Do they successfully create value for a diversity of stakeholders (Mastelic et al., 2015) or scale solutions from local experiment to broader regime changes (Von Wirth et al., 2019; Paskaleva & Cooper, 2021)?

The body of literature on Living Labs hardly involves critical evaluations on their performance, or their research design (Veeckman et al., 2013; Paskaleva & Cooper, 2021). This is caused by a variety of factors including a lack of practical evaluation tools (Overdiek & Genova, 2021) and a lack of incentive for critical reflection by founding parties that for instance depend on public grants (Paskaleva & Cooper, 2021). The fact that Living Labs bring the different domains of public actors together with private companies, knowledge institutions and citizens make that evaluations are often oriented towards one domain (perspective and standards) of the initialising and funding party. They are thus evaluated as research projects, public policy instruments, industry innovation, or community development vehicles. Recently, researchers related to ENoLL suggested a new evaluation framework (Vervoort et al., 2022) drawing on three levels: macro- (network of different stakeholders that engages in knowledge transfers around an innovative infrastructure), meso- (innovation projects and activities carried out) and micro-level (methodological steps and tools). It will be interesting to observe how these criteria will (or will not) be adopted by the broader lab family.

We observe that the broader evaluation of the Living Lab methodology is complicated by the different objectives held by particular Living Labs: their experimentation goals (product, service, value-chain, system) and the different industries and contexts in which they are situated. As mentioned, Living Labs are increasingly popular to tackle complex societal problems. Instead of product, service, or even local social innovation, this involves the transformation of socio-technical systems (Ceschin & Gaziulusoy, 2016). The fields of system innovation or transition studies (such as Strategic Niche Management and Transition Management) and (applied) design research thus increasingly crossover (Joore & Brezet, 2015), using comparable terminology (labs) while being rooted in different theoretical traditions and analytical focus points. For instance, in the field of transition studies, experimentation environments do not necessarily involve co-creation (Sengers et al., 2016) while this is seen as a core element of user-centred and participatory design inspired Living Labs (Broek et al., 2020). When evaluating Living Labs, differences like these should be considered and weighted against the proclaimed goals of the particular Living Lab.

Goals of Living Labs

When we zoom in closer on the different goals of Living Labs as places and projects, two major polarities come into focus: that between techno/managerial and social/civic goals, and that between aiming at (market-oriented) innovations or wanting to contribute to societal transitions. In this paragraph we would like to elaborate on these distinctions, as we think they reveal an important field of collaboration between Living Labs with different goals and orientation for further development. We have argued that the current practice of Living Labs is mainly oriented towards the innovation of different markets and geographies. However, transition theory and management thinking has been around for the past 25 years. Sengers et al. (2019) show in their systematic review on transition experiments that these experiments (mainly conducted as scientific research projects) are embedded in different logics and normative goals.

Another prominent polarity in the practice of Living Labs shows in the increased attention for transition. Living Labs increasingly pronounce their dedication to foster societal transitions, like that to a circular economy, a fossil-free energy system or alternative food and health systems, resonating with a transition focus in recent government policies (Het Groene Brein, 2021). Transition studies address transitions as long processes (40-80 years) of societal change. One popular way to look at transitions is Geels' multi-level perspective (2002; 2011) where societal macro structures (belief systems, paradigms) are related to existing institutions (meso or 'regime' level) as well as to micro practices and activities of different societal actors. It is in this micro field that so-called 'niche' actors are experimenting with new technologies and new ways of working that could possibly replace the old system. Whereas innovation does not necessarily have to change regime structures or system paradigms, transitions need to do so. This is also why it will probably not be one radical innovative product or service which changes a system. However, if several radically novel products, services, and technical systems are connected (across experiments) and combined with social innovations, transition might get into sight.

Hyysalo et al. (2019) have pinpointed that it is the long-range pathway focus of transition management (next 40-80 years) that often hinders more concrete experiments to be enabled and connected to each other. They suggest more mid-range pathway creation (next 10-20 years) related to experimental projects. Mid-range pathways could strengthen Living Labs, as concrete experiments in which stakeholders can experience and learn about possible futures as their core contribution. We as authors of this overview chapter on experimental environments find that, in practice, many Living Labs are still using methods for industry sector (food, logistics, fashion, etc.), geographical (city, region) or governance innovation with novel products, services, or technical systems and are not equipped to collaborate on systemic transitions in the way they are working. However, we see a trend and necessity in addressing more systemic questions in Living Labs, which influences the dynamic field in which Living Labs develop. We will elaborate on this idea further in the next chapter.

Innovation versus Transition Goals

Many labs will still be necessary to work on innovation goals, whereas other labs or lab programs may be more broadly geared to specific societal transitions. Arguably, when aiming for transition, this always involves innovation, but when a Living Lab is geared towards transition, the challenge is to safeguard that innovation is indeed embedded into a larger societal change. This will require a different methodological embedding of experiments and learning in Living Labs and their possible effects on society. In Table 1 we make some suggestions to this point.

Table 1. Innovation versus transition focus in Living Labs.

Living Labs that aim for (product, service, technical system) innovation	**Living Labs that want to contribute to (system) transition**
Can be short-lived, introduced by one larger project	Require long-term perspective, multiple projects
User-driven, technology-driven, market-oriented	Driven by future scenarios, necessities of larger social systems (earth climate, biodiversity, healthy living, etc.), values
Offer (isolated) solutions on different levels (product, service, product-service system, bounded socio-technical system)	Holistic perspective, understanding complex relationships and structures of societal systems
Can but do not necessarily influence regimes	Require breakdown of old systemic structures, regimes

Williams & Robinson (2020) made a first attempt for an evaluation framework for sustainability transition experiments. They draw on three aspects: evaluation of the process (e.g., inclusiveness and transparency), societal effects, and the broader impact on sustainability transitions. Because of its complexity and non-linearity, the last remains a domain for much more research. However, several authors remark on the mutual reinforcements and dynamics between experiments. Catering to these insights, we suggest an adaptation of the Living Lab family circle (Figure 1) in order to build a language for labs to situate themselves towards each other and develop more collaborations on common transition goals. For this, we add the dimension innovation (below) and transition (above) as a y-axis to the circle. Figure 2 shows how this transforms the family circle from Figure 1 to a canvas, which could be used by lab initiatives to make methodology choices transparent and relate labs to each other, thus using the variety of Living Labs for subsequent experimentation goals. Figure 3 shows what this could look like for the case of subsequent experiments shifting the production-consumption system with biobased products.

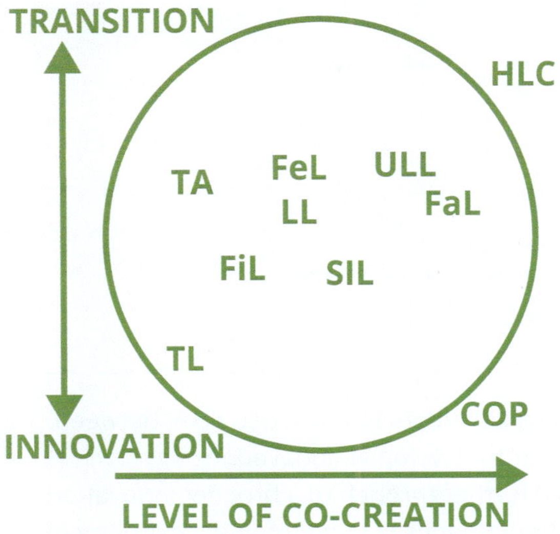

Figure 2. Matrix connecting experimental environments.

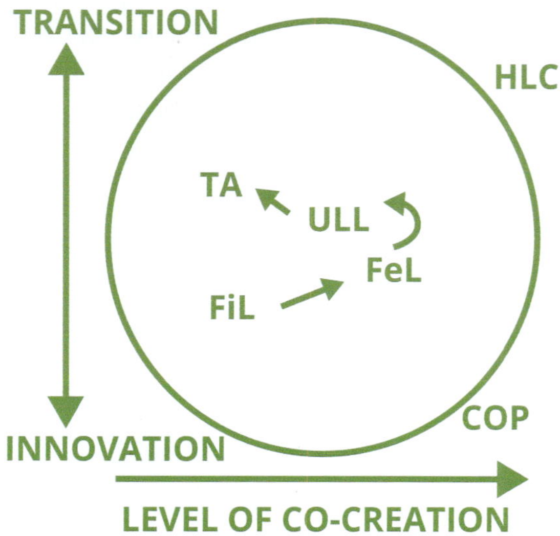

Figure 3. Matrix filled in for the case of biobased production.

The matrix could help stakeholders to make decisions between different types of labs, or to develop an innovation (e.g., regenerative materials) further through to a broader innovation or transition, using different types of experimental environments to get there. As experiments with technology can usually be done in shorter time frames than social experiments, the matrix could also guide stakeholders to experimental environments that offer a short-term exposure to users and systems (like Festival Labs) or a longer exposure within particular communities (like neighbourhood embedded ULLs).

From Designing Solutions to Shifting Systems

We believe that design and Applied Design Research (ADR) keep being crucial for taking ideas and methods not only to the theory, but also into the practice of real-life experimental environments, aimed at either innovation or transition. We will elaborate on this thought in the next section, before finalising this chapter with critical directions for the development of Living Labs, while arguing that the role of designers is especially needed to deal with these in the future.

Designers and design researchers have been involved in Living Labs and other experimental environments from the start, given the capacities they bring in for social and technical innovation. These design capacities can be summarisedas follows (Aguirre et al., 2017; Tonkinwise, 2015; Ceschin & Gaziulusoy, 2016; Sjöman & Hesselgren, 2020; Drew et al., 2021; Van Arkel & Tromp, 2023; Tromp & Vial, 2023):

- Visualisation or propositionality
- Prototyping or probing
- Integrative thinking
- Reframing
- Creative making
- Orchestration

While design qualities as these are deemed crucial in solving complex societal challenges, we have now reached a point in time that no longer involves solving problems, but (co)-designing completely new systems (Sjöman & Hesselgren, 2020; Jones & Van Ael, 2022; Zivkovic, 2018). The past thirty years, the design discipline has indeed expanded from designing sustainable products and services to designing the conditions in which these very things are used: processes, strategies, and visionary futures (Ceschin & Gaziulusoy, 2016). In summary, the same shift we observe in Living Lab practice can be found in the broader design practice. What does this mean for the future role of designers?

Design theory is self-critical about providing superficial solutions and facilitating incremental changes and insufficiently addressing long term, systemic challenges in the past (Ceschin & Gaziulusoy, 2016; Drew et al., 2021). Design practised with the goal of changing a system, coined system-shifting design (Drew et al., 2021) or transition design (Tonkinwise, 2015), goes beyond design practised with awareness of the system context it is embedded in. The UK Design Council argues that designing for transition requires a move (1) from a human-centred to a collective focus (stakeholders, ecosystems, and non-humans), (2) from designing out risk to transformative design, (3) from offering solutions to inviting new possibilities, and (4) from static solutions to designing for dynamic conditions (UK Design Council, 2021).

Firstly, a collective focus goes beyond human-centred design; it encompasses the 'more than human', all living organisms and materials. It asks designers to care for the system as a whole (Drew et al., 2021). Secondly, transformative design intentionally raises the question of what could be and entails imagining conditions in which transformation can emerge. It unlocks possibility by continuously asking, 'if this, then what else?'. Thirdly, design that is not solution-focused but 'stays with the trouble' (Haraway, 2010) is never done; it offers a springboard that others can build on. Finally and relatedly, system-shifting design means designing for longer-term processes of change, operating at each level of the system, including the systemic level of paradigms and collective narratives.

The four described characteristics of system-shifting design overlap with the distinction between Living Labs aimed at transition versus those aimed at innovation (Table 1). They mark an emergent practice in design, observed by the UK Design Council during the COVID-19 pandemic. As we argue, it is important to be aware of starting assumptions when setting up a Living Lab; it is equally important for designers to position themselves towards the dynamic between designing for innovation or transition. This offers a base against which outcomes can be evaluated, and a mental framework that guides thinking during the design process. Deliberately choosing system-shifting design as strategy, designers are required to break free from mental frameworks, to recognize that they themselves are embedded in the social structures and social systems they wish to change (Vink, 2021), and to motivate other stakeholders to do so as well. While we argue that current Living Lab practice is often not equipped to contribute to transitions, future design practice can play a role in employing system-shifting strategies to accomplish more disruption, and systemic transformation through labs.

Directions of Further Development

To finalise the chapter, we would like to point to some directions for the development and application of Living Labs and other environments for innovations, transitions, and collaborative multi-stakeholder learning.

A transparent and applied culture of co-creation and learning

The actual level of co-creation in Living Labs has been contested by different claims and expectations. Participatory design and co-design (Ehn, 1993; Sanders & Stappers, 2008; Blythe 2004, 2008) emphasise co-creation as a collaborative decision-making process with active involvement of users and stakeholders throughout the design process; whereas co-creation from a business perspective focuses on collaboration between customers and solution providers in the implementation phase. Governance perspectives have for a long time worked with different levels of co-creation: following Arnstein's 'participation ladder', they ranked citizen involvement based on their power to act in decision-making processes (e.g., citizens being informed versus citizens being equal partners) (Tritter & McCallum, 2006).

These differences in framing the meaning of co-creation led to very different ideas and methods of participation in the practical process of co-creation in Living Labs. Moreover, researchers question whether the level of co-creation proclaimed by the initiators of a Living Lab is actually delivered (Steen & Van Bueren, 2017; Greve et al., 2017; Huang & Thomas, 2021), which is problematic given that a high level of co-creation between all partners is one of the main promises of Living Labs (Maas et al., 2017). Designers and Applied Design Research with their knowledge around co-design and inclusion, and their interest in enabling multi-stakeholder boundary-crossing could be of great help in developing sharable and transparent methods of co-creation and evaluation thereof for experimental environments.

Frameworks and methods for the durable impact of labs

More than half of Living Labs registered at ENoLL are short-lived (Ballon & Schuurman, 2015) and the average research project using Living Labs is four years. As we have fleshed out by Table 1, when Living Labs are geared towards transition, a long-term perspective is imperative. Literature suggests four strategies to increase durability of an experiment: growth, replication, circulation, and institutionalisation (Broek et al., 2020). Living Lab environments which are permanent places to facilitate experiments need to guide projects more in deciding on and building such a strategy. As listed in the last chapter, designers and ADR have capacities to support labs in finding methods and designing tools for this. To increase the legitimization of Living Labs as a methodology for transition, Living Labs also need an approach that is designed to increase long-term involvement, replication, and relevance. Several design researchers have already pleaded for the design of frameworks for the 'hand-over' of learnings from one lab to another, and for the design of networks that transcend locations or connect locations (Ballon & Schuurman, 2015; Mastelic et al. 2015; Bergema & De Lille, 2022).

Creativity, methods, and spaces for discontinuous change and learning

Although Living Labs, and equally so, designers, are often seen as a catalyst for radical innovation (Sjöman & Hesselgren, 2020), this hinges upon their efforts to facilitate disruption or discontinuous change. Complex societal problems, such as those requiring system transitions, are known for their interconnected nature; they ask for an equally large multitude of perspectives to fathom. This is why Living Labs aimed at transition need to make a continuous effort to include perspectives that challenge the status quo. Designers and imagination are needed, as in practice, it appears very difficult to find novelty and disruption in collaboration with stakeholders rooted in the 'here and now' (Brons et al., 2022), including parties funding the lab.

Designers are already applying inquiry and imagination in Living Labs (Steen, 2013). We argued in this chapter that designers that position themselves towards transition and co-design with various actors are in a unique position to bring in methods to responsibly release 'old' perceptions, roles, and practices, open up space for imagination, or take a stance to challenge existing structures.

List of Abbreviations

- COP – Community of Practice
- FaL – Fablab
- FeL – Festival Lab
- FiL – Fieldlab
- HLC – Hybrid Learning Community
- LL – Living Lab
- SIL – Social Innovation Lab
- TA – Transition Arena
- TL – Testlab
- ULL – Urban Living Lab

Bibliography

- Aguirre, M., Agudelo, N., & Romm, J. (2017). Design facilitation as emerging practice: Analyzing how designers support multi-stakeholder co-creation. *She Ji: The Journal of Design, Economics, and Innovation*, 3(3), 198–209.
- Alavi, H. S., Lalanne, D., & Rogers, Y. (2020). The five strands of Living Lab: A literature study of the evolution of Living Lab concepts in HCI. *ACM Transactions on Computer-Human Interaction (TOCHI)*, 27(2), 1–26.
- Ballon, P., & Schuurman, D. (2015). Living Labs: Concepts, tools and cases. *Info*, 17, 1–11.
- Bergema, K., & De Lille, C. S. H. (2022). Scaling Up: From labs to systemic change. *Proceedings of Relating Systems Thinking and Design, RSD11*. https://rsdsymposium.org/scaling-up-from-labs-to-systemic-change/
- Broek, J. van den, Elzakker, I. van, Maas, T., & Deuten, J. (2020). *Voorbij lokaal enthousiasme: lessen voor de opschaling van Living Labs*. Den Haag.
- Brons, A., van der Gaast, K., Awuh, H., Jansma, J. E., Segreto, C., & Wertheim-Heck, S. (2022). A tale of two labs: Rethinking urban Living Labs for advancing citizen engagement in food system transformations. *Cities*, 123, 103552. https://doi.org/10.1016/j.cities.2021.103552
- Ceschin, F., & Gaziulusoy, I. (2016). Evolution of design for sustainability: From product design to design for system innovations and transitions. *Design Studies*, 47, 118–163.

- Chesbrough, H. (2003). *Open Innovation: The New Imperative for Creating and Profiting from Technology*. Harvard Business School Press.
- Dell'Era, C., & Landoni, P. (2014). Living Lab: A methodology between user-centred design and participatory design. *Creativity and Innovation Management*, 23(2), 137–154.
- Drew, C., Robinson, C., & Winhall, J. (2021). *System-shifting design: An emerging practice explored*. Retrieved February 25, 2024, from https://www.designcouncil.org.uk/resources/guide/download-our-systems-shifting-design-report
- Ehn, P. (1993). Scandinavian Design: On Participation and Skill. In D. Schuler & A. Namioka (Eds.), *Participatory Design: Principles and Practices* (pp. 41–77). Lawrence Erlbaum Associates.
- Geels, F. W. (2002). Technological transitions as evolutionary reconfiguration processes: A multi-level perspective and a case-study. *Research Policy*, 31(8–9), 1257–1274. https://doi.org/10.1016/S0048-7333(02)00062-8
- Geels, F. W. (2011). The multi-level perspective on sustainability transitions: Responses to seven criticisms. *Environmental Innovation and Societal Transitions*, 1(1), 24–40.
- Greve, K., Martinez, V., & Neely, A. (2017). *Bridging the Co-creation Gap Between Co-creators, Companies and Living Lab*. Cambridge Service Alliance.
- Greve, K., Vita, R. D., Leminen, S., & Westerlund, M. (2021). 38 Living Labs: From niche to mainstream innovation management. *Sustainability*, 13(2), 791. https://doi.org/10.3390/su13020791
- Haraway, D. (2010). When species meet: Staying with the trouble. Environment and Planning. D, *Society & Space*, 28(1), 53–55. https://doi.org/10.1068/d2706wsh
- Het Groene Brein. (2021). *Houvast voor duurzame vernieuwers. Vier perspectieven op transitie denken en doen* [Whitepaper]. Retrieved February 25, 2024, from https://hetgroenebrein.nl/publicatie-whitepaper-houvast-voor-duurzame-vernieuwers-vierperspectieven-op-transitiedenken-en-doen/
- Huang, J. H., & Thomas, E. (2021). A Review of Living Lab Research and Methods for User Involvement. *Technology Innovation Management Review*, 11(9–10), 88–107.

- Hyysalo, S., Marttila, T., Perikangas, S., & Auvinen, K. (2019). Codesign for transitions governance: A mid-range pathway creation toolset for accelerating sociotechnical change. *Design Studies*, 63, 181–203.
- Jones, P. (2014). Systemic Design Principles for Complex Social Systems. In G. Metcalf (Ed.), *Social Systems and Design, Volume 1 of the Translational Systems Science Series* (pp. 91–128). Springer.
- Jones, P., & van Ael, K. (2022). *Design Journeys Through Complex Systems: Practice Tools for Systemic Design*. BIS Publishers.
- Joore, P. (2010). *New to Improve, The Mutual Influence Between New Products and Societal Change Processes* (PhD Thesis). Technical University of Delft.
- Joore, P., & Brezet, H. (2015). A Multilevel Design Model: The mutual relationship between product-service system development and societal change processes. *Journal of Cleaner Production*, 97, 92–105. https://doi.org/10.1016/j.jclepro.2014.06.043
- Kieboom, M. (2014). *Lab Matters: Challenging the practice of social innovation laboratories*. Kennisland.
- Leminen, S., Westerlund, M., & Nyström, A. G. (2012). Living Labs as open-innovation networks. *Technology Innovation Management Review*, 2(9), 6–11.
- Loorbach, D., & Rotmans, J. (2010). The practice of transition management: Examples and lessons from four distinct cases. *Futures*, 42(3), 237–246.
- Maas, T., van den Broek, J., & Deuten, J. (2017). *Living Labs in Nederland – Van open testfaciliteit tot levend lab*. Rathenau Instituut.
- Mastelic, J., Sahakian, M., & Bonazzi, R. (2015). How to keep a Living Lab alive? *Info*, 17(4), 12–25. https://doi.org/10.1108/info-01-2015-0012
- Norman, D. A., & Draper, S. W. (Eds., 1985). *User Centered System Design: New Perspectives on Human-Computer Interaction*. Lawrence Erlbaum Associates.
- Notermans, V. I., von Wirth, T., & Loorbach D. (2022). *An experiential guide for Transition Arenas*. DRIFT, Erasmus University Rotterdam, Rotterdam, the Netherlands.

- Overdiek, A., & Geerts, H. (2023). *Innovating with Labs 2.0. Creating space for sustainability transitions*. The Hague University of Applied Sciences.
- Overdiek, A., & Genova, M. (2021). *Evaluating living labs? An overview of existing methods and tools*. White Paper, The Hague University of Applied Sciences.
- Paskaleva, D. K., & Cooper, D. I. (2021). Are Living Labs effective? Exploring the evidence. *Technovation*, 106 (May), 102311. https://doi.org/10.1016/j.technovation.2021.102311
- Pierson, J., & Lievens, B. (2005). Configuring Living Labs for a 'thick' understanding of innovation. In *Ethnographic Praxis in Industry Conference Proceedings* (Vol. 2005, No. 1, pp. 114–127). Blackwell Publishing Ltd.
- Sanders, E. B.-N., & Stappers, P. J. (2008). Co-creation and the new landscapes of design. *CoDesig*n, 4(1), 5–18. https://doi.org/10.1080/15710880701875068
- Sanoff, H. (1990). *Participatory Design: Theory & Techniques*. Bookmasters.
- Sengers, F., Wieczorek, A. J., & Raven, R. (2019). Experimenting for sustainability transitions: A systematic literature review. *Technological Forecasting and Social Change*, 145, 153–164. https://doi.org/10.1016/j.techfore.2016.08.031
- Sjöman, M., & Hesselgren, M. (2020). Designerly Living Labs: Earlystage exploration of future sustainable concepts. In *Synergy-DRS International Conference 2020* (Vol. 2, pp. 787–802). https://doi.org/10.21606/drs.2020.307
- Steen, M. (2013). Co-design as a process of joint inquiry and imagination. *Design issues*, 29(2), 16–28.
- Steen, K., & van Bueren, E. (2017). The Defining Characteristics of Urban Living Labs. *Technology Innovation Management Review*, 7(7), 21–33. http://timreview.ca/article/1088
- Tiesinga H., & Berkhout, R. (Eds., 2014). *Labcraft. How social labs cultivate change through innovation and collaboration*. Labcraft Publishing.
- Tonkinwise, C. (2015). Design for transitions from and to what? *Design Philosophy Papers*, 13(1), 85–92. https://doi.org/10.1080/14487136.2015.1085686

- Tritter, J. Q., & McCallum, A. (2006). The snakes and ladders of user involvement: Moving beyond Arnstein. *Health Policy* (Amsterdam, Netherlands), 76(2), 156–168. https://doi.org/10.1016/j. healthpol.2005.05.008
- Tromp, N., & Vial, S. (2023). Five components of social design: A unified framework to support research and practice. *The Design Journal*, 26(2), 210–228.
- Van Arkel, T., & Tromp, N. (2023). *Een ontwerpende aanpak voor maatschappelijke opgaven: waar hebben we het over?* Discussion paper. Retrieved via https://research.tudelft.nl/en/publications/een-ontwerpende-aanpak-voor-maatschappelijke-opgaven-waarhebben
- Vervoort, K., Trousse, B., Desole, M., Bamidis, P., Konstantinidis, E., Santonen, T., Petsani, D., Servais, D., De Boer, D., Spagnoli, F., Onur, O., & Bertolin, J. (2022). Harmonizing the evaluation of Living Labs: A standardized evaluation framework. In L. Bitetti, I. Bitran, S. Conn, J. Fishburn, E. Huizingh, M. Torkkeli, & J. Yang (Eds.), *Proceedings of the XXXIII ISPIM Innovation Conference*.
- Vink, J., Wetter-Edman, K., & Koskela-Huotari, K. (2021). Designerly approaches for catalyzing change in social systems: A social structures approach. *She Ji The Journal of Design Economics and Innovation*, 7(2), 242–261. https://doi.org/10.1016/j.sheji.2020.12.004
- Von Hippel, E. (2005). Democratizing innovation: The evolving phenomenon of user innovation. *Journal für Betriebswirtschaft*, 55(1), 63–78. https://doi.org/10.1007/s11301-004-0002-8
- Voytenko, Y., McCormick, K., Evans, J., & Schliwa, G. (2016). Urban Living Labs for sustainability and low carbon cities in Europe: Towards a research agenda. *Journal of Cleaner Production*, 123, 45–54. https://doi.org/10.1016/j.jclepro.2015.08.053
- Westley, F., Goebey, S., & Robinson, K. (2017). Change lab/design lab for social innovation. *Annual Review of Policy Design*, 5(1), 1–20.
- Williams, S., & Robinson, J. (2020). Measuring sustainability: An evaluation framework for sustainability transition experiments. *Environmental Science & Policy*, 103, 58–66.
- Zivkovic, S. (2018). Systemic innovation labs: A lab for wicked problems. *Social Enterprise Journal*, 14(3), 348–366.

Part 1: Living Labs and Societal Transitions

Jeroen van den Eijnde,
Masi Mohammadi

3. The Art of Connection

Beyond the Borders of Safe Zones: a Living Lab Case Study

DOI: 10.1201/9781003491484

> *People look happy, lab places look creative and inspiring, the brochures show people effortlessly working together, workshop card decks look graphically splendid and inviting, the social innovation jargon radiates a sense of co-creation and endless possibilities, and some designs illustrate almost surreal oversimplification of social reality. (...) The reality is that the world outside the happy lab is merciless. (Kieboom, 2014, p. 31)*

Living Labs: a Wild Concept

Living Labs operate in a dynamic and multifaceted reality. They significantly differ from a scientific laboratory in various aspects. In a scientific laboratory, strict guidelines are followed, with controllable variables and predictable outcomes. Reality is simplified to a limited number of identifiable and controllable variables, creating a safe and structured research environment.

In contrast, a Living Lab represents a specific real-world context without clear guidelines or predictable outcomes. Researchers in this environment must navigate a multitude of interacting and often uncontrollable (f)actors. This lack of clear rules and controllable variables introduces a sense of uncertainty. This uncertainty is amplified by the complex nature of the problems tackled in Living Labs. These problems, known as 'wicked problems', often lack predefined goals, outcomes, or solutions and involve a wide array of stakeholders with diverse values, norms, and interests, making these challenges particularly daunting. Moreover, the reality that a Living Lab operates within everyday life's complexities adds another layer of uncertainty.

As Kieboom (2014) advises, it's important to avoid oversimplifying this dynamic and multifaceted everyday reality where Living Lab research and experimentation occur. Participants in a Living Lab, therefore, must navigate these uncertainties, trust in each other's perspectives, and be willing to step out of their comfort zones to collaborate within this unique environment.

In this article, we explore the 'Art of Connection' Living Lab based in Arnhem, Netherlands, structured into six distinct parts. The first section, 'Collaborative Dynamics: initiators, ownership, and roles within the AoC Living Lab,' sets the stage by discussing the foundational elements and the diverse stakeholders involved. Next, 'Financing models' examines the impact of funding on the collaborative environment. The narrative then moves to 'Collaboration and Co-creation,' analysing the interactions between partners and the process of joint innovation. 'Citizens in the Living Lab' emphasises the role of community engagement and citizen science in enhancing the lab's relevance and impact. The development and community response to the 'Prototype of the Mobile Interactive Artwork' is detailed in the fifth section. 'Student involvement in the AoC Living Lab' highlights the contributions of students, illustrating the educational benefits and new insights they provide. The article culminates in sharing key insights and practical lessons on fostering safe, responsible, and effective collaborations within complex Living Lab settings, while highlighting the critical role of citizen science in enhancing community engagement and outcomes (Lessons Learned and Key Findings).

Art of Connection Living Lab

The Art of Connection (AoC) Living Lab seamlessly integrates with the concept of a 'social innovation lab', as proposed by Kieboom (2014), effectively addressing the challenges associated with ageing. This initiative transcends traditional care frameworks, operating within an experimental environment to foster the creation of interactive artworks through collaborative efforts with end-users and stakeholders. This process is aimed at enhancing community familiarity in residential areas, supporting environments that facilitate light social interactions and frequent engagements among residents, thereby fostering a sense of belonging and security, and reducing isolation among older adults. The research, initiated in September 2021 and concluded in July 2024, was funded with approximately €300,000 by the Dutch Taskforce for Applied Research SIA. This investment positioned the project within the larger context of the Knowledge & Innovation Agenda for Health Care 2020-2030 (Zie literatuurlijst), demonstrating the Netherlands' dedication to evolving a mission-driven healthcare approach. This initiative aimed at markedly increasing the provision of care within personal living environments by 2030, shifting away from traditional long-term care facilities.

The primary goal of the AoC project is to enhance community familiarity and social interaction within neighbourhoods, which are pivotal in developing a supportive and caring environment. This is particularly significant in addressing social isolation among older individuals in the Netherlands, a condition that not only escalates health risks but also augments dependency on long-term care services. By reinforcing informal networks and social connections, the AoC project supports the independence of older individuals, addressing the challenges of double ageing, marked by an increase in the number of older individuals and their average life expectancy (Projectplan, 2020). While the initial project plan was highly praised by the SIA review committee for its innovative approach in engaging residents and focusing on a specific neighbourhood, concerns were raised about the engagement of 'end-users and vulnerable individuals' within the Living Lab setting and the project's potential impact on this group (Subsidieverleningsbesluit, 2021). In response, the research team facilitated the participation of approximately 130 older individuals, ensuring that the project more effectively incorporated these groups. This approach significantly enhanced the potential for impactful outcomes, aligning with the project's goal of fostering inclusive and supportive community environments.

Collaborative dynamics: initiators, ownership, and roles within the AoC Living Lab

The Art of Connection(AoC) research proposal, following a citizen-led initiative called CPO Cohousing, was developed by the Architecture in Health research group at HAN University of Applied Sciences. It adheres to the content and financial guidelines set forth by the aforementioned SIAcall. This research group engaged in initial discussions with local entrepreneurs and the care network in the Coehoorn neighbourhood of the city of Arnhem and was responsible for establishing the project's organisational framework. Notably, due to socio-economic factors, the originally planned neighbourhood of Coehoorn was not feasible for the project's execution. Consequently, an engaged group of residents from another Arnhem neighbourhood, Stadseiland, undertook the continuation of the project. Within this context, additional community groups, such as Impian Kita (an active Indonesian community) and residents of Eilandstaete (a nursing home), significantly contributed to the formation of the AoC Living Lab project.

Moreover, the research team, beyond drafting the project proposal, actively sought the participation of social and industrial partners, and of other research institutions. As co-initiators and principal investigators, HAN holds substantive and financial accountability for the project (Projectplan, 2020). The course of action, content of the project, and the extent of each partner's contribution were, however, thoroughly discussed and unanimously agreed upon by all participating entities.

In the AoC project, alongside resident groups, four research groups from universities of applied sciences participate, complemented by creative and different care partners. These research groups – Architecture in Health at HAN University of Applied Sciences (lead), Tactical Design at ArtEZ University of the Arts, AI & Big Data at Fontys UAS, and Healthy Society at Windesheim UAS – each contribute their expertise in areas such as healthy living environments, creative co-creation, AI, and Big Data for societal innovations and promoting a healthy society.

Creative partners, including Fillip Studios and Energy Floors, enhance the project with their contributions to the design and development of the artwork. Additionally, care partners, notably Rijnstate Hospital and residential care organisations, aim to play a pivotal role in outpatient and community-based care and healthy ageing initiatives (such as Ouder Worden 2040). However, aligning with the AoC project's emphasis on preventive health, some care organisations have opted for a more supportive, advisory role in the background, while others, like the nursing home Eilandstaete, remain actively involved. This arrangement is a strategic decision to engage care organisations in the project's setup and evaluation without necessitating a significant time investment from them. This method underscores an effort to foster a balanced collaboration, with each partner contributing their unique strengths towards achieving the AoC Living Lab's overarching goals.

Financing models creating restrictive collaboration frameworks in Living Labs

The AoC Living Lab exemplifies the importance of collaboration and co-creation between public and private partners, which is crucial for achieving the stated social and societal goals. However, like many Living Labs in the Netherlands, this lab relies on short-term government financing as a research project. These kinds of short-term funding mechanisms lead to the disappearance of experimental environments within a few years due to a lack of vision and follow-up resources (Exceed, 2023). This SIA subsidy, intended to promote applied research, requires an in-kind and/or cash co-financing of at least 50% of the total project costs by the project partners. A minimum of 75% of the subsidy must be allocated to the salary costs of the involved universities of applied sciences, which results in most of the work of the non-academic project partners being in-kind. The motivation of these partners to participate in such projects varies. For the SMEpartners of the AoC Living Lab, such as Fillip Studios and Energy Floors, developing a business case to contribute to reducing healthcare costs and expanding their market share is a key motivation (Projectplan, 2020). The healthcare organisations involved, given the project's focus on prevention and a community-based approach, play a modest role with limited co-financing (approximately 6%). Finally, CPO Cohousing, as the driving force behind the Living Lab, contributes almost a quarter of the total in-kind co-financing.

While the government's support has been instrumental in the project's initiation and development, the experiences and outcomes of AoC underscore the limitations of relying solely on public funding. Particularly, the challenges of sustaining innovative environments beyond the initial funding period highlight the potential benefits of a more diversified, hybrid financing model (Boual & Zadra-Veil, 2018; Gualandi & Romme, 2019). This approach would not only enhance the resilience and sustainability of projects like AoC but also enable a broader scope of experimentation and community engagement. In light of this, it is arguable that future projects should consider adopting hybrid funding structures. By blending government support with private and NGO contributions, Living Labs can secure a more stable and versatile foundation. This fusion encourages greater engagement from various sectors, leading to richer, more sustainable innovations that continue to thrive beyond the constraints of short-term government grants. Ultimately, the transition towards hybrid funding models could ensure that initiatives like the AoC Living Lab evolve into enduring fixtures within their communities, continuing to address societal challenges with innovative solutions.

Regarding the financing of the AoC Living Lab, these significant matters stand out. Firstly, there is a substantial investment in hours by knowledge and creative partners, which carries the risk of asymmetric collaboration. Secondly, the call does not provide an opportunity to financially compensate citizens for their contributions to the project, as only organisations registered with the Chamber of Commerce are currently eligible for subsidies. For the significant societal challenges addressed within open Living Labs, this can result in restrictive frameworks for collaboration with citizens.

Collaboration and co-creation

With the specific aim of collaboratively designing and developing a prototype for an interactive outdoor space within a defined timeline, the AoC project exemplifies a collaborative approach within the quadruple helix framework. This approach brings together partners from research, industry, healthcare, and the community to address complex societal challenges. The project's foundation on shared, cross-disciplinary ambitions is essential for tackling contemporary issues effectively. The need is evident for transdisciplinary collaboration, where knowledge from different disciplines is not just combined (multidisciplinary) or integrated (interdisciplinary) but also leads to new insights and applications across different kinds of knowledge. This is crucial in a Living Lab context, where innovation is jointly driven by researchers, firms, users, governments, and citizens (Nicolescu, 2008; Bernstein, 2015; Maas et al., 2017). Successful cooperation in such complex projects depends on various factors, including a clear shared goal, mutual trust and respect for each other's expertise, and effective communication (Schuurman et al., 2015; Leminen et al., 2012; Ståhlbröst, 2012). These factors are further supported by an environment where participants feel comfortable sharing their knowledge, experiences, and ideas.

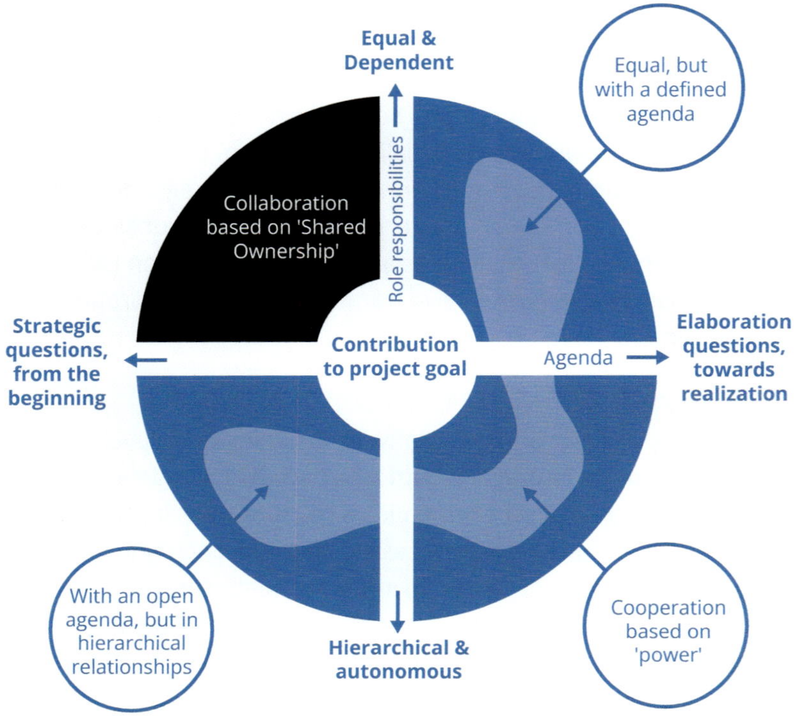

Figure 1. Forms of collaboration according to the quadrant model of Lambregts & Schipper, 2015 (source: Lambregts & Schipper, 2015, p. 19).

The Lambregts & Schipper (2015) model offers a valuable lens for examining the collaborative dynamics within the AoC project (Figure 1). At its core, the project is initiated by partners including the neighbourhood residents and the research group Architecture in Health at HAN University, embodying a collaboration rooted in shared ownership. In contrast, other partners engage under the principle of 'equality but with a defined agenda', highlighting the nuanced diversity in involvement and perception of roles within the project. This setup underscores the varying degrees of commitment and objectives among the participants, contributing to the project's rich collaborative landscape.

Aiming to design and develop a prototype for social interaction in the neighbourhood, the AoC project navigates between fostering creative flexibility (open outcomes) and meeting specific, predetermined goals (closed outcomes), achieving a blend of innovation and precise, targeted results in its collaborative efforts.

Shared ownership and engagement are critical to the success of collaborative projects like Living Labs. In AoC these were evidenced by active contributions and a sense of joint responsibility among partners, who brought their individual knowledge, skills, and experiences to bear. Lencioni (2002) outlines five characteristics of effective teams—focus on collective outcomes, initiative and responsibility, commitment, openness in conflicts, and trust—which were observed in the AoC team dynamics. These principles underpinned an equitable collaboration, allowing every partner to have a voice in the process, despite differences in roles and tasks. Supporting this collaborative ethos were both structural and behavioural components. Structurally, the project included strategy development, process steering, and program management to direct the collaboration's execution. The core team steered on equitable and dialogue-based collaboration. The leadership style within this project also fostered open dialogue among partners. The project leader actively engaged in resolving any ambiguities or misunderstandings through personal outreach, directly addressing issues related to collaboration, content, or differing interpretations. This hands-on approach promoted communication but also cultivated an environment of transparency and mutual respect.

Despite these intentions, the AoC project faced some challenges. Communication in such interdisciplinary projects remains sensitive and prone to misunderstanding. The challenges in communication, exacerbated by staff changes and the need for flexibility in response to socio-economic changes, highlight the complexity of managing such innovative projects. Although the Sway environment and regular meetings of the Project Management Team played a crucial role in facilitating communication and cohesion among the diverse partners, in hindsight, communication within this project could perhaps have been handled more effectively. In addressing these challenges, particularly those amplified by staff changes, and the constraints of the COVID-19 period, appointing a dedicated person for project communication could enhance clarity and efficiency. For instance, by designating a specific individual as the project communication officer, there would be a clear point of contact for all partners. This role could have mitigated some of the communication barriers experienced, ensuring consistent messaging, facilitating the flow of information, and addressing misunderstandings promptly.

Behaviourally, a focus on open dialogue among partners fostered a shared understanding, crucial for co-ownership. Project Management Team meetings, marked by robust dialogue, highlighted ongoing collaboration, challenges, and future directions. Learning from each other, or learning evaluation, was key to fostering continuous improvement.

While acknowledging the essential role of dialogue on political and ethical dimensions of national policies, including healthcare, and underscoring the importance of such deliberations in preventing 'political blind spots' as described by Kieboom (2014), it is paramount that such discussions are conducted during the project proposal phase. These early-stage dialogues enable a comprehensive understanding of the technological solutions and the dynamics of public-private partnerships, enriching the project's foundation. However, once the project is approved and underway, the focus must shift towards executing the agreed goals and experiments. This strategic timing ensures that while critical issues are addressed upfront, the project's momentum is maintained, allowing the team to concentrate on delivering the work as promised, without diversion or delay.

Despite the initial agreements detailed in the project plan, ongoing discussions among some partners about their contributions highlighted the challenge of aligning individual expectations with collective ownership. This situation reflects the broader complexities of collaborative efforts, underscoring the necessity for shared ownership, active participation, and clear communication. It also brings to light the divergence between objective agreements and subjective perceptions, a divergence that is deeply rooted in the dynamics of human relationships and the interpretation of work processes. Although roles and involvement were explicitly outlined in the project plan, the evolving nature of endeavours sometimes requires reconciling pre-agreed upon terms with emerging expectations.

The primary emphasis on user involvement in design research required management flexibility and continuous strategic goal alignment within the innovation process. This theme was pervasive throughout the AoC project, blending theoretical models with practical execution and underscoring continuous improvement, critical reflection, and the achievement and flexible adaptation of predefined goals.

Citizens in the Living Lab

The AoC Living Lab, situated in the Stadseiland neighbourhood and aimed at fostering public familiarity among its residents, is distinguished by its 'open approach'. This strategy, however, introduced significant challenges due to the complexity and diversity of the target audience. In our research, we initiated a series of activities designed to gather the views, needs, and preferences of the neighbourhood residents (Figures 2, 3, 4, and 5). The details of these activities are outlined below. Overall, more than 130 residents participated in this project.

Figure 2. Researchers conducting in-depth dialogues with an actively engaged focus group in the participants' homes.

Neighbourhood-oriented product design: navigating the broader community landscape

In Living Labs, end-users are regarded as the foremost experts. This project featured a notably committed group of residents who not only participated actively from the outset but also served as co-initiators, constituting the project's core group. Their dedication and engagement proved to be invaluable, even though they represented only a segment of the neighbourhood community. When addressing neighbourhood-oriented product development, issues such as representation are crucial, impacting sustainable use and overall acceptance. Consequently, a pressing question surfaces: how can one ensure collaboration with a group of residents that truly mirrors the diversity of the neighbourhood?

The challenge of representativeness in the context of Living Labs and participatory design is widely recognized in the literature. Sanders & Stappers (2008) highlight the importance of involving a diversity of users in the design process to arrive at holistic solutions. Yet, this ideal of complete representativeness often clashes with the constraints of time and resources, especially in projects aimed at rapid product development like AoC. Therefore, the project opted for a strategic approach by working intensively with an active core group of elderly residents while also seeking broader citizen support through fairs and information evenings. This layered approach found backing in the literature, with Bratteteig & Wagner (2012) emphasising the value of diverse participation strategies for inclusivity in design processes.

Figure 3. Attendance at meetings organised by residents to understand the neighbourhood's needs and to introduce the project plan.

Neighbourhood fair

Existing literature emphasises the importance of public space events, such as fairs, in fostering community engagement and gathering data through participatory design and action research (e.g., McNiff & Whitehead, 2009). Aligning with this perspective, a fair was organised in the Stadseiland neighbourhood by the research team and the active focus group on a Saturday. The event aimed to introduce the project at the neighbourhood level, gather feedback on conceptual plans for the artwork, and, importantly, enhance interaction and community spirit. Featuring a diverse program that included performances by the neighbourhood band and culinary delights such as 'saté from Impian Kita,' the fair attracted approximately 80 residents from a variety of backgrounds and age groups. This gathering fostered a welcoming atmosphere essential for stimulating open dialogue and reinforcing community bonds. The residents' positive feedback highlighted broad support for the project and underlined the significance of such events in both soliciting feedback and nurturing a sense of community.

Several activities, including an Augmented Reality presentation of the proposed prototype and feedback sessions, provided preliminary insights into the residents' preferences regarding the interactive artwork and their eagerness to connect within the community. These endeavours underscored the residents' desire to actively shape their living environment and established the fair as a foundational platform for participatory research. The event revealed the necessity for better communication strategies and clearer information, especially about the project's adaptability and the artwork's location. Despite the initial nature of these insights and the highlighted need for more in-depth information gathering, the fair's success and constructive feedback illustrated the value of community events in obtaining critical insights for the project's iterative development.

It's crucial to acknowledge that a few neighbourhood residents strongly opposed installing any product in the neighbourhood's open spaces, which they view as their communal backyard. This resistance showcased the community's diverse perspectives and the essential need for inclusive engagement and dialogue in the participatory design process. Recognizing and addressing these concerns is key to building community ownership and aligning the project with the interests and expectations of all neighbourhood residents. This dynamic underscores the pivotal role such events play in fostering community engagement, collecting valuable data, and navigating the complexities of achieving community consensus and giving room to controversy within projects like AoC.

Addressing these viewpoints led to a significant shift in the project's direction—from a stationary concept to developing a more mobile artwork. This change not only demonstrates a responsive and adaptable approach to community feedback but also prepares the ground for discussing the evolved prototype concept in the subsequent section.

Figure 4. A fair method to introduce the project at the neighbourhood level involves exploring the broader community landscape and gathering feedback on the artwork through targeted approaches, thereby enhancing interaction and fostering community spirit.

Neighbourhood-oriented product development: methodology in the AoC Living Lab

In Living Labs, particularly given their open nature, the Empathic Design Framework emerges as a crucial tool with its systematic approach to exploring, translating, processing, and validating. This framework streamlines innovation by ensuring that techno-spatial solutions are deeply rooted in the lived experiences and implicit needs of users (Mohammadi, 2017). In the AoC Living Lab, the initial phase was dedicated to exploring the needs and preferences of neighbourhood residents through a variety of qualitative research techniques, including focus groups, interviews, and the creation of personas that represent average neighbourhood residents. These techniques, along with creative sessions, were utilised to gain insights into user needs and experiences, providing a solid knowledge base for the design phase.

Acknowledging the diversity within the neighbourhood and the challenges of engaging a variety of target groups, the AoC Living Lab employed methods, tailored to its limited scope, to engage a broader segment of the community. Different data collection methods—including walking interviews, reflection sessions, a fair, and attendance at meetings organised by residents—facilitated dialogue with the community, aiming to gather preliminary insights within the constrained timeframe. In stark contrast, our engagement with the highly motivated focus group enabled a deeper exploration and collaboration, yielding a more nuanced understanding of user experiences and expectations. This approach underscores the project's dual strategy: efficiently navigating the broader community landscape with targeted methods, while achieving profound insights through close collaboration with a dedicated group of participants.

The gained insights were considered in the historical and demographic context of the Stadseiland neighbourhood and then translated into a design brief, which artists used to design concept products. Cultural probes and Augmented Reality (AR) techniques utilised during a fair and focus group meetings used to evaluate these concepts. This approach fostered primary community engagement and feedback on the intended prototypes. This feedback and the findings were integrated into an adapted prototype version for planned validation in the neighbourhood. The current version of the prototype will be discussed in the next section.

Figure 5. Eilandstaete co-creation: collage created by residents about their ideas on social interaction in the neighbourhood.

Prototype of the mobile interactive artwork

Exploring the intersection of art, technology, and community engagement, we delve into the development of a mobile interactive artwork intended for neighbourhood integration. The realisation of the prototype and its introduction to the neighbourhood are forthcoming steps. The development process of the artwork included two design iterations. In the first design round, a digital prototype was created, employing AR to digitally experience and test the artwork. During this phase, residents were closely involved and could provide valuable feedback on the digital prototype during various meetings and even at the local fair. The feedback, as described in the paragraph 'Citizens in the Living Lab', was crucial and led to the evolution of the initial concept, transforming the original idea of fixed neighbourhood furniture into a versatile 'mobile meeting and inspiration table' in the form of a cargo bike. Although the product is still under development, figure 6 presents an image of the current design.

However, it's important to note that a significant validation phase in a real-life setting in the neighbourhood is still to be done. Practical issues, such as maintenance and storage, have been discussed with the residents' group.

Figure 6. Preliminary design of a mobile and interactive meeting place (design and realisation by Fillip Studios Arnhem)

Student Involvement in the AOC Living Lab

From HAN, Fontys, and ArtEZ, a total of 28 BA Product Design students contributed to this project in 2023 and 2024. For example, in the ArtEZ educational project, students collected information from residents using cultural probes (Figure 7). In this project, a student approached residents in a shopping centre, asking them to mark on a map where they felt safe or unsafe. Another project engaged residents in creating a visual story about their favourite animal and place in the neighbourhood. This story was then made available on a website for further contributions. The project 'Do You Wanna Be Baking Friends?' invited residents to bake bread using provided dough, following their cultural traditions, and share their experiences. This initiative fostered interaction among some neighbourhood residents. More than forty residents provided information in this manner, and the students received feedback from some residents and a design agency.

Figure 7. ArtEZ students BA Product Design offer their cultural probes to local shoppers. (Photo credit: Anna Fehn and Jelle Zegers)

Lessons learned and key findings

The AoC project exemplifies the core principles of collaboration and realisation within the framework of social and urban innovation Living Labs, through an exploration of co-creation and participatory methods. This project has offered insights into the dynamics of developing inclusive solutions for societal challenges, facilitated by the cooperation among a diverse array of stakeholders, including residents, researchers, healthcare providers, students, and professionals from the creative industries. This interdisciplinary approach has led to the creation of tangible prototypes and interventions that enhance community engagement in neighbourhoods, while also deepening our understanding of the complex social structures and diversity within these areas.

What unites all partners in this project is their deep commitment and shared sense of societal importance and urgency to increase social engagement in the neighbourhood and to demonstrate the societal added value of collaboration in a quadruple helix.

The study highlights the vital importance of sustaining ongoing and constructive dialogue among all project partners, tackling challenges related to collaboration, financing, and project management. Such interactions demand a high degree of flexibility, transparency, the fostering of mutual trust, and the development of a shared sense of project ownership. Even in scenarios where the contributions of time and financial resources among Living Lab participants vary, it is crucial to establish clear expectations regarding everyone's roles, motivations, and aspirations from the start of the project. In the AoC project, all partners were provided with and consented to an initial research plan, which detailed the financial contributions expected from each participant. At the outset of the project, partners were also given the chance to express their individual viewpoints on the goals and methodologies, as previously outlined in the research plan, and to introduce ongoing and relevant programs in which they were involved. This procedure aimed to cultivate a unified vision and foster effective collaboration. Nevertheless, it is understood that perceptions of this process's effectiveness may vary among partners. In the realm of interdisciplinary collaboration within Living Labs, the necessity for ongoing improvement in communication is evident, particularly in response to changes. Tailoring communication strategies to these changes is essential for keeping partners aligned and guaranteeing the project's ongoing success.

The success of Living Lab projects is critically dependent on their ability to adapt flexibly to dynamic and evolving circumstances. Equally vital is their ability to actively incorporate the community's needs into the development process. The AoC project serves as a prime example of such adaptability and resilience, particularly in response to unforeseen challenges, including the COVID-19 pandemic, modifications in the case study, and shifts within team and organisational structures. These adjustments underscore the essential conditions required to successfully navigate the complexities inherent in social innovation processes. Furthermore, the role of funding bodies, exemplified by SIA in this context, is paramount. Their readiness to accommodate changes—as demonstrated by adjustments in partnerships, timeline extensions, and financial modifications—proves to be critical. This level of flexibility is not just supportive but essential, as it allows Living Labs to effectively tackle the dynamic and unpredictable nature of real-world challenges, ensuring that the integration of the community's needs remains central to the innovation process.

Moreover, the significant engagement of an active group of residents underscores the impact of targeted participation and the value of personal commitment in achieving substantial project results. This element of the research illuminates the capacity of relatively small but highly motivated groups to serve as catalysts within broader community endeavours. The utilisation of co-creation methods, as demonstrated by the researchers, student-led projects, and the neighbourhood fair, has fostered resident involvement and highlighted the role of art in promoting social cohesion and public engagement.

In conclusion, the AoC project offers lessons for the development of future Living Lab initiatives, highlighting the critical need for flexibility, active participation, and interdisciplinary collaboration to foster social innovation. Moreover, with its community focus, it presents a stark contrast to traditional healthcare models, emphasising the imperative for innovative ways of health improvement by social interaction. Additionally, it underlines the significance of integrating art and architecture in the transition towards community-based care, showcasing how these elements can significantly contribute to enhancing the effectiveness and appeal of such care models.

Bibliography

- Bernstein, J. (2015). *Transdisciplinarity: A Review of Its Origins, Development, and Current Issues*. Journal of Research Practice, Volume 11, Issue 1, Article R1
- Bratteteig, T., & Wagner, I. (2012). *Disentangling power and decision-making in participatory design. Proceedings of the 12th Participatory Design Conference: Research Papers – Volume 1.* https://doi.org/10.1145/2347635.2347642
- Boual, J., & Zadra-Veil, C. (2018). *New hybrid organizations in the social and solidarity economy in France: Providing public goods and commons*. https://doi.org/10.25518/ciriec.css1chap13.
- Bowker, G. C., Timmermans, S., Clarke, A. E., & Balka, E. (Eds.). (2015). *Boundary objects and beyond: Working with Leigh Starr*. Cambridge/London: The MIT Press.
- Broek, J. van den, Elzakker, I. van, Maas, T., & Deuten, J. (2020). *Voorbij lokaal enthousiasme – Lessen voor de opschaling van Living Labs*. Den Haag: Rathenau Instituut.
- Call for proposals Praktijkgericht onderzoek voor extramurale zorg. (2020). Utrecht: Nationaal Regieorgaan Prakijkgericht Onderzoek SIA (onderdeel van NWO) & Health Holland.
- Clark, A. J. (2018). *Design Anthropology: Object Cultures in Transition*. London: Bloomsbury Academic.
- Dell'Era, C., & Landoni, P. (2014). *Living Lab: A Methodology between User-Centered Design and Participation Design. Creativity and Innovation Management, 23*(2), 137-154. https://doi.org/10.1111/caim.12061
- EXCEED: Expert Coalition for Experimental Environment Development. (2023). *CLICKNL, Topconsortium voor Kennis en Innovatie van de Topsector Creatieve Industrie*.
- Georges, A., Baccarne, B., Schuurman, D., & Coorevits, L. (2015). *User engagement in Living Lab field trials. Info*. https://doi.org/10.1108/info-01-2015-0011
- Greve, K. Vita R. De, Leminen, S., & Westerlund, M. (2021). *Living Labs: From Niche to Mainstream Innovation Management. Sustainability, 13*, 791, 1-25. http://doi.org/103390//su13020791
- Gualandi, E., & Romme, A. (2019). *How to Make Living Labs More Financially Sustainable? Case Studies in Italy and the Netherlands. Engineering Management Research*. https://doi.org/10.5539/EMR.V8N1P11.

- Gunn, W., Otto, T., & Smith, R. C. (Eds.). (2013). *Design Anthropology: Theory and Practice*. London: Bloomsbury Academic.
- Hong Huang, J., & Thomas, E. (2021). *A Review of Living Lab Research and Methods for User Involvement*. *Technology Innovation Management Review, 11*(9/10), 88-107.
- *KIA: Kennis en Innovatieagenda Gezondheid & Zorg 2020-2030 (2019)*. Vitaal functionerende burgers in een gezonde economie. Health Holland.
- Kieboom, M. (2014). *Lab Matters: Challenging the practice of social innovation laboratories*. Amsterdam: Kennisland.
- Lambregts, P., Schipper, L. (2015). *Gedeeld eigenaarschap. Nieuw perspectief op samen werken aan maatschappelijke veranderopgaven*. Vakmedianet.
- Leminen, S., Westerlund, M., & Nyström, A.-G. (2012). *Living Labs as Open-Innovation Networks*. *Technology Innovation Management Review, 2*(9), 6-11.
- Lencioni, P. (2002). *'The Five Dysfunctions of a Team: A Leadership Fable.'* , Hoboken, NJ: Jossey-Bass Publishers.
- Maas, T., Broek, J. van den, & Deuten, J. (2017). *Living lab in Nederland – Van open testfaciliteit tot levend lab*. Den Haag: Rathenau Instituut.
- Mastelic, J., Sahakian, M., & Bonazzi, R. (2015). *How to keep a Living Lab alive?*. *Info, 17*(4), 12-25. http://dx.doi.org/10.1108/info-01-2015-0012
- McNiff, J., & Whitehead, J. (2009). *Doing and Writing Action Research*. Thousand Oaks, CA: Sage.
- Mohammadi, M. (2017). *Empathische woonomgeving*. Technische Universiteit Eindhoven.
- Nicolescu, B. (2008). *Transdisciplinarity: Theory & Practice*. Cresskill: Hampton Press.
- *Ouder Worden 2040. Een transformatieagenda voor een ouder wordende samenleving* (2022).
- Paskaleva, K., & Cooper, I. (2021). *Are Living Labs effective? Exploring the evidence*. *Technovation, 106*, 1-10. http://doi.org/10.1016/j.technovation.2021.102311
- *Projectplan The Art of Connection* (2020). Arnhem: HAN Lectoraat Architecture in Health.

- Sanders, E.B.N., & Stappers, P.J. (2008). *Co-creation and the new landscapes of design.* CoDesign, 4(1), 5–18. https://doi.org/10.1080/15710880701875068
- Schuurman, D., et al. (2015). *Living Labs: A Systematic Literature Review. Open Living Lab Days 2015 Conference Proceedings*.
- Smith, R. C., Vangkilde, K. T., Kjaersgaard, M. G., Otto, T., Halse, J., & Binder, T. (Eds.). (2016). *Design Anthropological Futures*. London: Bloomsbury Academic.
- Ståhlbröst, A. (2012). *A Set of Key Principles to Assess the Impact of Living Labs*. *International Journal of Product Development, 17*(1/2), 60-75.
- *Subsidieverleningsbesluit 2021 inzake aanvraag subsidie Praktijkgericht onderzoek voor extramurale zorg, indieningsronde november 2020, EXZ.EXZ.01.009 'The Art of Connection'*. Utrecht: Regieorgaan SIA, 1 juni 2021.

Janneke Sluijs, Maria Arias, Morgan Duta,
Ju Laclau Massaglia, Anja Overdiek

4. Co-designing towards Transitions?

Facilitating more than Experiment in Living Labs

DOI: 10.1201/9781003491484

Due to their unique mindset and skills, designers are particularly apt to facilitate transformation processes in Living Labs, especially those that have social innovation as a goal. However, existing literature geared to the practice field runs short when designers interested in taking on such facilitation roles search for concrete orientation on how to best tackle that challenge.

Future-Proof Labs within the Designing Value Networks research group of The Hague University of Applied Sciences carried out research on different multi-stakeholder co-creation cases, between 2017 and 2024. The cases were aimed at developing experiments and ultimately shifting systems in the context of sustainability transition efforts. While experiments tackled both design of technical and social innovations, we will concentrate in this chapter on the importance of using Living Lab projects to work specifically on social innovations, and on the role of the design facilitator in this. We will propose a framework that can act as a lens through which the different stakeholders can understand and relate their collaborative activities to the larger system shift they aim for. Based on this, we also present first results on a facilitation approach and a set of skills that can support designers who aim to tackle such facilitation tasks.

Introduction

Living Labs are experimental environments that allow different stakeholders to collaborate in the exploration of new approaches to contribute to societal transitions to more sustainable practices in various areas, which can range from energy (e.g., that societies need to move away from fossil fuels and towards more sustainable energy sources), food (e.g., taking steps away from animal proteins through the adoption of non-animal alternatives or making production and consumption more local), healthcare, and education. Due to more traditional thinking about the innovation power of markets, people in these fields still tend to focus on designing and launching different products and services to the market (Loorbach et al., 2017; Ceschin & Gaziulusoy, 2016). These products and services, however, typically imply changes in value chains and other ways stakeholders are relating to each other and working together. They cannot grow within the current socio-technical systems and therefore do not automatically alter dominant practices.

If we want to transform these dominant practices, we need Living Labs that not only allow for experimentation on the product and services level, but also on the systemic level. As explained by Loorbach et al. (2017), transitions are non-linear and can last decades, usually between 40 and 80 years, since they require changes to be made on multiple levels and a shift in systemic perspectives. This includes social innovation, which refers to the social transformations needed for societal transitions like new relationships, behaviours, and attitudes (Bock, 2012) between, e.g., value chain partners, citizens, and municipalities or private companies and governments. Design scholars (Hyysalo et al. 2019; Gaziulusoy & Ryan, 2017) have argued that experiments need to be based on more mid-term pathways, which range between 5 and 15 years, and need to be related to the whole spectrum of design (Lähteenoja et al., 2023). Experiments are required to move beyond the creation of new products and services, so as to allow for and include uncertainty, the use of new language, new relationships among stakeholders, and learning and reflecting across systems. Additionally, all stakeholders need to be brought 'into the lab' to allow true co-creation to take place.

Designers are particularly apt to facilitate such multi-stakeholder co-creation processes due to their so-called designer mindset (e.g., Howard et al., 2015), which includes the ability to work in uncertainty, to creatively reframe common approaches to reality, to visualise complex information and to use empathy for understanding social situations (Design Council, 2021a; Tromp & Vial, 2023). When facilitating multi-stakeholder lab collaborations in practice, designers are often approached by an enabling institution (typically a Public-Private Cooperation or a municipality) to explore the situation among the stakeholders of a certain societal challenge, bring them together, and help them align around future visions and possibly shared tasks and goals. Establishing dialogue among stakeholders is a key component of the early stages of a Living Lab process (Yasuoka et al., 2018). Scholars at the intersection of transition studies and design (Botero et al., 2020) suggest the term 'translation' for the work of navigating between mundane practices of stakeholders, design strategies, design process facilitation, and design outcomes. While the translation concept is interesting to explore further, we will stick to the facilitation task in this chapter. Translation as a task still remains abstract, and our goal is to inform design practitioners with the concrete aim of facilitating Living Labs focused on the social innovation aspects necessary for societal transitions.

Hence, we looked at the facilitation role of the designer in projects geared towards transitions through the lens of facilitating social innovation with a designer mindset. In this chapter, we propose that designers and design methods are wellsuited to take the lead in this 'more than experiment' methodology and that designers can co-design with multi-stakeholder networks to achieve meaningful transformations. Additionally, we propose a framework and explore the skills that designers need to use when facilitating 'more than experiment' in the Living Lab.

More Than Experiment – the Changing Role of the Designer

Living Labs are built upon a methodology that focuses on co-creating knowledge and developing a collective imagination. When looking at the insights yielded by the human-centred design field, particularly in the Human-Computer Interaction community, it is clear that user-centred development and testing of design solutions in Living Labs leads to successful design outputs, both in the form of new products and services (Alavi et al., 2020). Here, the output of a Living Lab is framed as a technological innovation. Even if it is looked at as a socio-technical innovation, changes in mindsets or relationships between the stakeholders are often not captured as results.

In contrast to this, an emergent practice is design facilitation, which involves applying design processes and approaches to enable dialogue and ideation in participatory design contexts to develop new solutions to complex problems (Mosely et al., 2021). The literature on this practice focuses either on the facilitator's characteristics and behaviour or on the characteristics of the joint activity and the environment in which it is carried out. Aguirre et al. (2017) explain that the designers' role in co-design workshops counts with three main characteristics: providing creativity and imagination, making abstract tasks tangible and experienceable, and bringing the user into the collaboration as a stakeholder. Additionally, according to Akama & Light (2020), aside from leading participants through the specific activities performed during the course of one or several workshops), the designer also tends to keep an eye on the overall journey to guide the group to the next steps in the 'right' direction. However, formal methods alone are not enough when faced with the inherent uncertainty of a co-design process aimed towards social innovation. Thus, designers need to prepare themselves for the in-situ facilitation.

The literature, however, describes little about the unexpected dynamics that unfold while facilitating dialogues and sessions with the stakeholders. Authors such as Granholt & Martensen (2021) and Mosely et al. (2021) have expressed the need for both studies that explore the performative nature and creative process of design facilitation and also for further research on the skills and tools needed to facilitate sessions that guide non-designers through the design process. Furthermore, Mosely et al. (2021) identified that designers act as facilitators in two instances: during the course of specific co-design workshops and as part of the broader project implementation, which shows that the role of the designer goes beyond the experimentation and implementation phases of a Living Lab project.

Design facilitators' constant shifting between the session level and the overall project level is also noted by other experts. The role of designers in Living Labs have been described by the Design Council (2021a) as 'convenor', 'communicator', 'maker', and 'systems thinker or integrator'. A convenor is 'someone who has good relationships, can create spaces where people from different backgrounds come together, and joins the dots to create a bigger movement' (Design Council, 2021a) and therefore needs to watch out for complexity and competition when working with multi-stakeholder groups. Thus, designers are working more and more beyond the experimentation phase across the whole project leading to a Living Lab and capturing its results. This process involves co-designing with diverse networks and catalysing systemic change. For instance, social designers are doing this by using 'resilience-driven design activities', which shift the focus from the user to society by addressing issues that most of its members are affected by with the aim of improving their shared future (Tromp & Vial, 2023).

It is important to note that, according to transition theory, niche experiments are essential to feed societal transitions (Geels, 2002;

Geels, 2006; Van Buuren & Loorbach, 2009; Williams & Robinson, 2020) as they can lead to new ways of thinking and working. Niche experiments are often linked to the development of new social networks and innovative technologies, which can expand, change the system, and develop the power to influence larger systemic beliefs about the purpose of societal systems. However, in other cases, local experiments do not lead to the broader adoption of new technologies or work methods among stakeholders that would be necessary to allow for a bigger transition – like those of the energy or food sectors – to take place (Schuurman et al., 2016; Ballon et al., 2018; von Wirth et al., 2019). We suggest in this chapter that design facilitation with a focus on social innovation can strengthen the positive link between local experimentation and broader transitions.

The following section will introduce a framework developed to describe the Living Lab process, which was used to explore a facilitation approach and to find skills that can be useful for designers who want to contribute to social innovations. These findings are based on empirical data collected between 2017 and 2024. The ultimate goal will be to dive deeper into the question:

What approach and skills could support designers who want to facilitate co-design in Living Labs aimed towards social innovation within transitions?

The Framework

Over the past seven years, Future-Proof Labs, our Living Lab research, has explored and represented visually the process a multi-stakeholder network undergoes when experimenting in local labs. It has also contributed to Living Lab theory with four design principles and five basic rules for successful multi-stakeholder experiments in Living Labs (De Lille & Overdiek, 2021; Bergema & De Lille, 2022; Overdiek & Geerts, 2021; Overdiek & Geerts, 2023). It is important to note that we understand 'successful multi-stakeholder experiments' to be those that bring about the design of innovative technologies and social innovation in the form of multi-stakeholder learning, thus having a positive effect on the transition, even after the lab has ended. Therefore, under the right circumstances, Living Labs can lead the different stakeholders involved to creative and collaborative learning, developing shared technologies, and exploring new ways of thinking and working.

Based on research across Living Lab cases in the aviation and retail industries and those geared towards the energy and circular transitions (De Lille & Overdiek, 2021; Bergema & De Lille, 2022; Overdiek, 2024), we have developed the Impacting Systems with Local Experiment (ISLE) model to visualise how local experiments are situated within the broader Living Lab process and how learnings can be initiated at different moments on the way to systemic changes. The ISLE-model (Figure 1) visualises the ideal systemic innovation process from a current to a future (sub)system. When assessing the scale levels informed by the multi-level perspective on transitions (y-axis) over time (x-axis), the model shows how designers can co-design with multi-stakeholder groups towards systemic change. Three main phases are distinguished within this process: first, when stakeholders collaboratively understand the current system and make sense of desirable futures, then when they engage in local experiments together with (more) users of future designs, and finally when working design outcomes (social and technological) are redesigned and implemented towards preferred future systems.

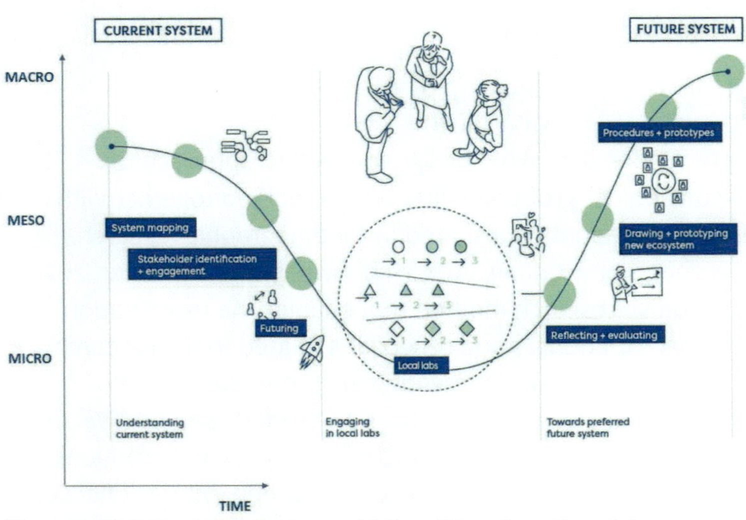

Figure 1. ISLE (Impacting Systems with Local Experiment) model (version from Overdiek, 2024).

Instead of intervening only at one point in the lab collaboration, such as when reframing the future vision together with stakeholders or conceiving and prototyping solutions together with users, designers have started to realise their broader role across the different phases of the systemic innovation process (Design

Council, 2021b; Drew et al., 2021; Leadbeater & Winhall, 2020). We used the ISLE model as a framework to think about the skills designers might need when facilitating 'more than experiments' in Living Labs.

Based on insights found through the literature review and our field research, the facilitation process is twofold. On the one hand, it involves facilitating individual workshops and interactions at different moments of the Living Labs process. This facilitation needs an approach that invites social innovation. On the other hand, the facilitation process requires navigating across the whole project, or even different similar processes aimed at making impact on the same systemic change. This insight led our research question to morph into two sub-questions:

1. *How can a co-design facilitator approach the multi-stakeholder group in the different moments of interaction, in order to invite social innovation?*
2. *What skills are needed to (a) navigate across the whole process and (b) facilitate interventions in the process?*

Methodology

Finding an approach for different moments of interaction

To better understand what designers need to approach the multi-stakeholder group, the first author worked together with a design practitioner on a shared autoethnographic research. Auto-ethnography (Chang, 2016; Schouwenberg & Kaethler, 2021) is a qualitative research method that enables the examination of personal experiences and reflections related to the research context. In the context of navigating interventions in multi-stakeholder transformation, it offers an insider's perspective and allows for the inner conditions, such as thoughts and emotions, to be recorded. It furthermore encourages reflexivity, a reflection on actions and decisions, and helps to uncover underlying assumptions and motivations. The data that was used consisted of personal written reflections, before and after sessions further described by the 'Case Background', verbal reflections through dialogues with involved stakeholders on their experience, and an ongoing weekly dialogue for the duration of 6 months between the researcher and the practitioner for self-inquiry on their inner conditions throughout the process.

The first three dialogues resulted in 6 hours of audio recording that have been used to capture the practitioner's knowledge. Subsequently, the transcription was analysed for emerging themes and translated into an artefact that represents the practitioner's knowledge. Since this knowledge is unique to this individual practitioner and is embedded in tacit understandings, the use of artefacts in research can help externalise it, as explained by Candy & Edmonds (2010).

Using research-through-design (Stappers & Giaccardi, 2017), this visual artefact was employed to plan and reflect on three sessions in a Living Lab (see 'Case Background'). The visual artefact and iterations of it, created after each session, informed the approach to facilitate the sessions with a multi-stakeholder group. In total, we created four artefacts. The first and second iterations of these artefacts and reflections together represent the findings of this research.

Case Background

A local Living Lab is being set up within Greenport West-Holland, a greenhouse horticulture innovation cluster, to achieve multi-stakeholder co-design activity between the Dutch greenhouse industry, a municipality, the regional government, and citizens. The Living Lab was to be focused on local circularity around green waste streams.

Since it was still in an early stage, it was suitable for an exploration of the dynamics of the 'understanding the current system' phase, as referred to by the ISLE model. The three facilitation sessions mentioned above had the intent of serving to identify stakeholders, engaging them, and gathering them around a shared purpose and experiment, which is a characteristic part of this stage.

Diving deeper into the skills

In order to answer the second research question about the skills needed to facilitate interventions within and navigate throughout the Living Lab process, a focus group was organised consisting of design practitioners with different degrees of experience. As part of the focus group sessions, participants were introduced to the ISLE model, which they used to describe the skills they would typically need as design facilitators in the different phases of a Living Lab project. This resulted in a preliminary set of 101 described skills, which were transformed into prototype cards.

The definition of 'skills' as a concept was researched further through a literature review on competencies and through three two-hour workshops with groups of four to twelve participants, including experienced design facilitators, practitioners, researchers, and lab project managers. Participants were asked to examine, reduce and categorise the 101 cards, as shown in Figure 2. This process led to a reduced set of 69 cards.

Three more workshops were organised with three different groups of fifteen experienced design practitioners to explore the potential of these cards and develop new iterations that could be more relevant and functional for design practitioners. Additionally, the workshops allowed the cards to be iterated with an eye on functionality, accessibility, and clarity.

Figure 2. Research partners examining skills cards.

The results of both the approach research and the skills research were analysed separately by the five authors of this chapter to minimise the influence of individual bias and then discussed together. Three collaborating practice partners (Van Waarde consultancy, Being A Designer, and Fundamentals Academy) and a researcher from the Expertise Network Systemic Co-design (ESC) were also involved in the discussion on a bimonthly basis, which added up to six times over a one-year period.

Results

Finding an approach for different moments of interaction

The auto-ethnographic analysis of the approach used by facilitators during different moments of the interaction yielded two main insights. The first insight is that to approach the multi-stakeholder group, an *understanding of how the perceived realities of stakeholders mediate and form the problem* is important. The second insight is that it is crucial for transformations to happen, to *facilitate an encounter with these different realities and to arrive at synergy with the group.* For this to be effective, *different levels of awareness* need to be designed for. An important insight here is that between the perceived realities and abstract value goals of individual actors are these levels of awareness. We concluded that it is important for the facilitator to understand how these different levels of awareness could and should be accounted for and how co-design activities could engage them, which is further explored through artefact 1 (Figure 3).

Navigating between multiple realities

During the facilitation process, the designer needs to navigate between the concrete problems presented by the stakeholders, which can be described as the gap between the reality they perceive and what they would like reality to look like, as well as the more abstract societal values and goals related to the broader transition goal. The left half of the first artefact (Figure 3) depicts a range of levels from earthly matters, the experience of reality on the bottom to more abstract values on the top. Five colours are used in the background to represent different levels of human awareness which are relevant for facilitators to work with. These awareness levels start at the bottom with the belief system and continue towards the emotional system, the power-ego or 'I' level, the synergy or 'we' level, and the leadership level.

The awareness levels can be recognized from what is more common knowledge about human development through stages of physical, emotional, intellectual, and social development. However, only data from the auto-ethnographic research – and no additional theory – was used to conceive these levels. Using simple quotes (e.g., 'That's how it is', 'That's how you act', and 'What's good for the group is good for me'), the artefact depicts how these awareness levels might be represented in statements of the individual participant of the lab.

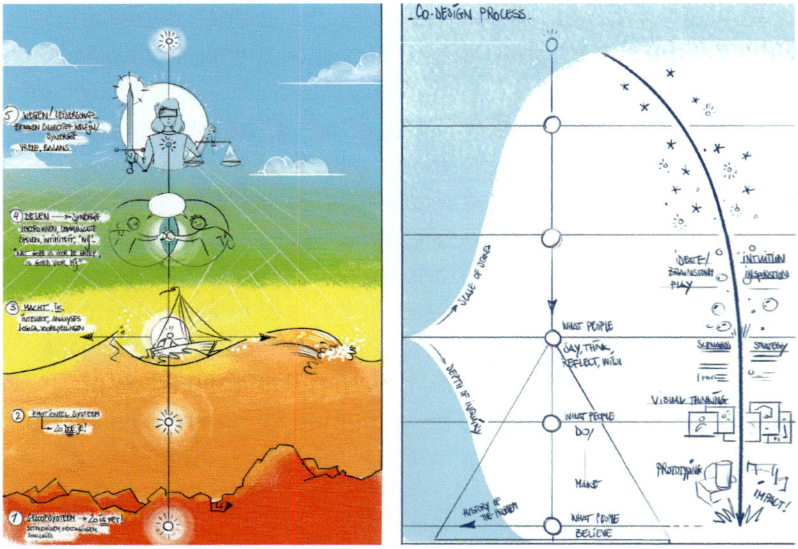

Figure 3. Visual artefact 1 that describes levels of awareness (left) as beliefs, emotions, rationality, synergy, and leadership and links these to activities and concepts from the co-design field (right).

The right half of the artefact depicts the co-design process, activities (e.g., visual thinking and prototyping) and outcomes (e.g., strategy and scenario) in a way that visually aligns them to the levels of awareness on the left. We draw on aspects which are common knowledge in co-design, such as the levels of knowledge which range from explicit (what people say and think) to tacit (how people do and use) and latent knowledge (what people believe).

During the first Living Lab session informed by this approach, the intent was a deeper exploration of the problem and the stakes in the form of perceived realities and meanings. The perceived realities were explored through interviews of the participants in pairs. Questions like 'What do you believe is the problem? What would you like to happen? And what happens if we don't do anything?' served as prompts for the participants to answer verbally and then draw, thus engaging them on the intellectual and emotional levels.

During the third Living Lab session, these topics were revisited. Participants were asked to introduce themselves through photographs that showed how the issue of green waste streams shows in their reality, which they were asked to prepare and bring to the session. Then, a round of interviews in pairs was conducted in which participants were asked to annotate these photos with the emotions they evoked. The annotated photos were used to form a big map depicting the perceived problem(s) through the eyes of all participants. An example from the session of how beliefs filter perceived realities is Figure 4. One participant labels this pile of green waste mixed with plantingpots as 'horrible', because of her belief that having the two streams mixed means that they both go to waste. Another participant with knowledge on the filter capacity of their green waste processes has labelled the same picture as 'fantastic'. This way of working, depicting concrete realities and exchanging how people give meaning to them, allowed the participants to encounter each other's perceived realities.

Building up towards group synergy

The second insight is that it is crucial for transformation to happen to *facilitate an encounter with these different realities and to arrive at synergy with the group.* For this to happen, *different levels of awareness* need to be designed for.

Figure 4. Picture annotated with contrasting emotions that were evoked. Depicted in the picture is a pile of green waste streams mixed with planting pots.

The artefact as a whole aligns the levels of awareness through which humans perceive and engage with the world with the co-design activities and levels. This artefact could help facilitators get oriented on co-design activities that engage the different awareness levels in participants. Synergy (level 4 in Figure 3) is not something that we can directly influence with co-design activities. Rather, it is an emergent property from the levels below that are being engaged in all participants through intentional co-design activities. The facilitator is then responsible for creating a dynamic balance through designing with the levels below in mind, allowing all participants individually and in the group a composure on each level. This supports the conditions conducive for group synergy to occur.

In the moments of interaction, the facilitator's role is to observe if synergy is an emergent property during the session by observing the levels below and to register any interferences. The interferences can show up on any of the levels. On the intellectual level, they show through unclear communication or misunderstandings which could arise from hidden assumptions, or conflicting values or ideas. On the emotional level, they are evident through emotions that go unnoticed or unacknowledged or those that are very dominant. On the physical level, they show through conditions that are distracting or uncomfortable. Guided by the artefact (Figure 3), this approach led to a rich stakeholder involvement by attending to the different levels of awareness through the intentional use of co-design methods.

The realisation of our perceived realities and engaging all levels of awareness resulting in group synergy are two elements of an emerging new facilitation approach we would like to propose for designers facilitating multi-stakeholder groups in labs. We call it the 'Designing Synergy' approach. More elements might be added to this approach once it has been applied in different contexts and across different phases of a Living Lab project.

Diving deeper into the skills

The first key insight about the skills needed to navigate co-design processes with multi-stakeholders towards transitions is related to the way we refer to said skills. During the first card-sorting workshops, with 101 previously determined skills, participants divided the cards into personal attributes (being) and skills (doing).

Certain attributes were perceived to be inherent to a person (i.e., curiosity) and may enable the execution of different skills (i.e., digs deeper), see also Figure 5. Based on this observation, a literature review on competency concepts revealed that the cards could better be distinguished as attitudes (A) being inherent to the self and knowledge and skills (KS) as something that can be acquired and trained. This is why we conceptualised the necessary skills as KSA.

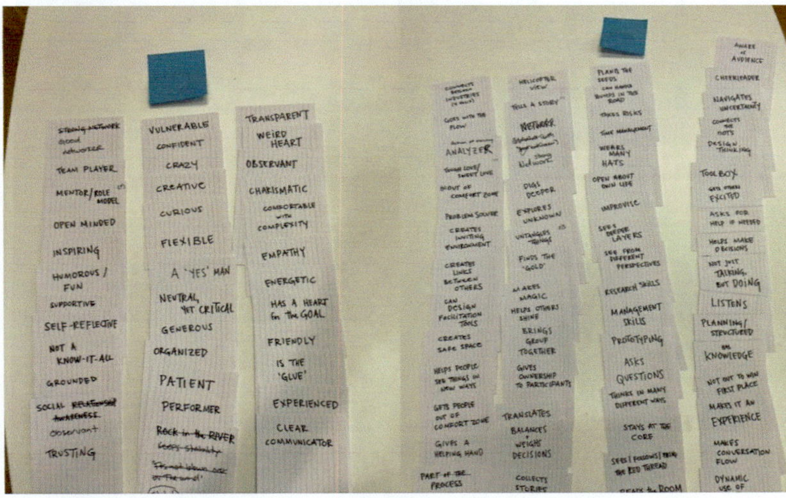

Figure 5. The original 101 skills sorted into two groups: 'Being' (left) and 'Doing' (right).

The second insight is that facilitating a single session requires different skills than navigating an entire Living Lab process geared to systemic change. The KSAs could also be sorted in this way, dividing them between those needed to facilitate a session (micro) and those needed to strategically think about and steer the process (macro). Therefore, the next iteration of the KSA cards was categorised into those two groups distinguished by card colour: the edge of the cards is grey for micro KSAs, and white for macro KSAs, as shown in Figures 6 and 7.

The third insight revealed that different KSAs had different levels of complexity. Therefore, a third category was distinguished from the two mentioned above and called the baseline group (as shown in Figure 8), since these KSAs were found to be generic skills of a trained designer. The baseline skills would often act as a foundation to develop micro facilitation skills, which in turn could become a foundation for more long-term and strategic macro skills.

Figure 6. KSA cards representing micro skills.

Figure 7. KSA cards representing macro skills.

Additionally, during the workshops, it was discussed that some KSAs within the micro and macro groups could be further clustered into groups (i.e., 'making sense for myself' or 'controlling the conversation' within the micro category). To help the clustering process, cards identified as belonging to the same group were colour-coded. The cards were found to spark discussions on the skills topic among the different groups of junior or more senior design facilitators, and the colours of the cards were found to help with this.

The three insights above and additional card-sorting led to 69 KSA cards describing skills needed to facilitate in and across the Living Lab process and work towards systemic change. We observed that these cards which use practitioners' language were employed by designers in different settings to talk about and determine skills they need when fulfilling these facilitation tasks. Thus, they could be a starting point for designers (and facilitation teams) who look for orientation to spark a conversation about the KSAs they find important and how to further develop them.

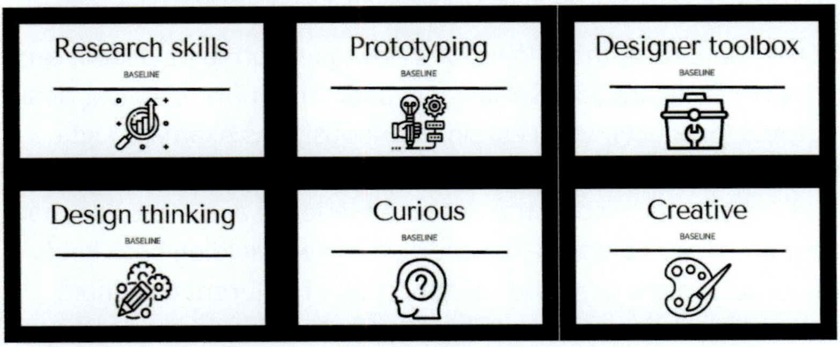

Figure 8. KSA cards representing baseline skills.

Discussion

Little is known about the relation between local experiments in labs and systemic transitions and, in particular, what designers who act as facilitators for such systemic change projects should know and do. Shifting our focus to the facilitation of sessions and the navigation of the overall project process related to social innovations yielded new insights into the knowledge and skills needed by designers who want to tackle such facilitation tasks.

Our ISLE framework combined with the Designing Synergy approach and the 69 KSA cards could work as a starting point for designers who look for orientation when preparing for and fulfilling such tasks. Additionally, by employing said knowledge and skills, designers could find more concrete ways to use their design and facilitation skills to reach new, broader levels of social innovation.

It is important to note, however, that the artefacts explored in this chapter – the ISLE model, the visual prototypes for the approach, and the cards – need to be further applied, researched, and developed across different contexts of systemic change.

Further research into the facilitation of social innovation should cause the social mechanisms needed for social transformation – such as new relationships, behaviours and attitudes – to be increasingly seen as design outputs, alongside products, services, and technical systems. As explained by Akama & Prendiville (2013), the way to accomplish this is through shifting away from the focus on methods and pre-designed proposals and step into the human 'in-between' space that is dynamic, emergent, and relational (Akama & Prendiville, 2013). Our 'Designing Synergy' approach is one way to prepare for this dynamic co-design practice.

At the same time, this allows for a new question to be posed: can a designer facilitate both social innovation and more 'classical' design outputs (products, services) simultaneously? As Akama & Light (2020) mention, design facilitators are only ready for what they are and are only as good as the design experience and expertise that they are 'warmed up to'. Therefore, a project yielding both kinds of outputs might need the collaboration of different designers with different experience and expertise. This would be certainly worthwhile to explore further.

What is apparent from our research is that to facilitate a multi-stakeholder group towards a transition, a designer needs to be able to go back and forth among different levels of abstraction: a more abstract one when navigating the Living Lab process geared to systemic change and a more concrete one when facilitating a single intervention (i.e., strategizing versus executing the strategy, macro KSAs versus micro KSAs). It is also important to keep in mind that, aside from the skills of the designer, the co-design itself is essential, which means that the knowledge and skills brought to the table by every stakeholder makes a difference. To explore these stakeholder skills would also be an interesting new avenue for design research in experimental environments.

Finally, we would like to address the idea that navigating a Living Lab project geared towards transition can be planned entirely or that a one-dimensional cause-and-effect relation can be drawn between design facilitation and societal transformation, or even the adoption of a new way of working after the end of a Living Lab project. Emergence as the self-developing property in complex systems remains an important force in the co-design process which can be used and reduced or amplified by the design facilitator, though further research should be done on how and in how far to channel this emergence in lab collaborations.

Conclusion

Based on earlier research into co-designing in Living Labs and further research conducted for this chapter, we can conclude that local co-design activity in multi-stakeholder Living Labs is a suitable environment to explore how stakeholders relate to each other and which behaviour and attitudes are needed to impact societal transformations. This chapter aimed to find ways to support facilitating designers as they tackle this task. To accomplish this, the chapter explored how the co-design facilitator can approach the multi-stakeholder group prior and during the interaction, and what knowledge and skills designers need for facilitating specific interventions and for navigating across the Living Lab projects geared to societal transitions.

On the one hand, when considering the facilitation approach, we described different levels of awareness that help bring into focus the social aspects while preparing for and helping throughout the facilitation process. Understanding these awareness levels can help the facilitator craft a dynamic balance through the use of co-design tools and activities, register interferences, and respond to them systemically to achieve synergy. However, it is worth noting that the proposed Designing Synergy approach has only been tested in western environments and during what the ISLE model refers to as the 'first phase of a transition project' (see Figure 1). It would seem that in the second phase, where the testing of new concepts together with users is prominent, a design output focus might be combined with a social innovation approach.

On the other hand, after an inventory of the necessary skills, we arrived at 69 KSAs (knowledge, skills, and attitudes) that designers facilitating social innovations need in order to facilitate more than experiments in Living Labs. We designed and iterated a set of 69 KSA cards depicting micro skills to conduct interventions and macro skills to strategically plan the next steps towards systemic change by looking through a macro lens. Additionally, competencies found to be inherent to the design discipline were placed into a baseline category of skills cards. Testing has shown that these cards can provide a 'shared language' to talk about skills for Living Lab projects geared towards societal transitions. The cards can help practitioners to engage in a dialogue on what facilitator skills are needed at different moments in a transition project or which ones are missing from a specific team to tackle an upcoming project. Further research could explore how far these cards can also help co-designers from diverse stakeholder groups to become more acquainted with collaborating in a Living Lab context.

Bibliography

- Aguirre, M., Agudelo, N., & Romm, J. (2017). Design facilitation as emerging practice: Analyzing how designers support multi-stakeholder co-creation. *She Ji: The Journal of Design, Economics, and Innovation, 3*(3), 198–209. https://doi.org/10.1016/j.sheji.2017.11.003
- Akama, Y., & Light, A. (2020). Readiness for contingency: Punctuation, Poise, and co-design. *CoDesign, 16*(1), 17–28. https://doi.org/10.1080/15710882.2020.1722177
- Akama, Y., & Prendiville, A. (2013). Embodying, enacting and entangling design: A phenomenological view to co-designing services. *Swedish Design Research Journal, 9*, 29–40. https://doi.org/10.3384/svid.2000-964x.13129
- Alavi, H. S., Lalanne, D., & Rogers, Y. (2020). The five strands of Living Lab. *ACM Transactions on Computer-Human Interaction (TOCHI), 27*(2), 1-26. https://doi.org/10.1145/3380958
- Ballon, P., van Hoed, M., & Schuurman, D. (2018). The effectiveness of involving users in digital innovation: Measuring the impact of Living Labs. *Telematics Informatics, 35*(5): 1201–1214. https://doi-org.hr.idm.oclc.org/10.1016/j.tele.2018.02.003
- Bergema, K., & De Lille, C.S.H. (2022). *Scaling Up: From Labs to Systemic Change. In Proceedings of Relating Systems Thinking and Design (RSD11)*. Brighton, UK, Oct. 3–16, 2022. https://rsdsymposium.org/wp-content/uploads/2023/01/RSD11_paper_121.pdf

- Bock, B. B. (2012). Social innovation and sustainability; how to disentangle the buzzword and its application in the field of agriculture and rural development. *Studies in Agricultural Economics, 114*(2), 57-63.
- Botero, A., Hyysalo, S., Kohtala, C., & Whalen, J. (2020). Getting participatory design done: From methods and choices to translation work across constituent domains. *International Journal of Design*, *14*(2), 17-34.
- Candy, L., & Edmonds, E. (2010) The role of the artefact and frameworks for practice-based research, in M. Biggs & H. Karlsson [Eds]. *The Routledge Companion to Research in the Arts*. Abingdon: Routledge, pp. 120–137.
- Ceschin, F., & Gaziulusoy, I. (2016). Evolution of design for sustainability: From product design to design for system innovations and transitions. *Design Studies*, *47*, 118-163.
- Chang, H. (2016). *Autoethnography as Method* (Vol. 1). Routledge.
- De Lille, C.S.H. & Overdiek, A. (2021). *From system to local to system: Design principles to scale for a system in transition. In Proceedings of Relating Systems Thinking and Design (RSD10)* 2021 Symposium. Delft, the Netherlands, Nov. 2-6, 2021. http://openresearch.ocadu.ca/id/eprint/3854/
- Design Council. (2021a). *Beyond Net Zero. A Systemic Approach*. https://www.designcouncil.org.uk/fileadmin/uploads/dc/Documents/Beyond%2520Net%2520Zero%2520-%2520A%2520Systemic%2520Design%2520Approach.pdf
- Design Council. (2021b). *System-Shifting Design. An Emerging Practice Explored*. https://www.designcouncil.org.uk/fileadmin/uploads/dc/Documents/Systemic%2520Design%2520Report.pdf
- Drew, C., Robinson, C., & Winhall, J. (2021). *System-Shifting Design an Emerging Practice Explored. In Proceedings of the Relating Systems Thinking and Design Symposium*. University of Brighton, Brighton, U.K., October 13-16. https://rsdsymposium.org/wp-content/uploads/2023/01/RSD11_paper_24.pdf
- Gaziulusoy, A. I., & Ryan, C. (2017). Roles of design in sustainability transitions projects: A case study of Visions and Pathways 2040 project from Australia. *Journal of Cleaner Production*, *162*, 1297-1307.

- Geels, F. W. (2002). Technological transitions as evolutionary reconfiguration processes: a multi-level perspective and a case-study. *Research Policy, 31*(8-9): 1257-1274. https://doi.org/10.1016/S0048-7333(02)00062-8
- Geels, F. W. (2006). Multi-level perspective on system innovation: Relevance for industrial transformation. *Environment & Policy*, 163–186. https://doi.org/10.1007/1-4020-4418-6_9
- Granholt, M., & Martensen, M. (2021). Facilitate design through improv: The qualified eclectic. *Thinking Skills and Creativity, 40*, 100785. https://doi.org/10.1016/j.tsc.2020.100785
- Howard, Z., Senova, M., & Melles, G. (2015). Exploring the role of mindset in design thinking: Implications for capability development and practice. *Journal of Design, Business & Society, 1*(2), 183-202.
- Hyysalo, S., Marttila, T., Perikangas, S., & Auvinen, K. (2019). Codesign for transitions governance: A mid-range pathway creation toolset for accelerating sociotechnical change. *Design Studies*, *63*, 181-203.
- Lähteenoja, S., Marttila, T., Gaziulusoy, İ., & Hyysalo, S. (2023). Transition co-design dynamics in high level policy processes. *Design Studies*, *88*, 101207.
- Leadbeater, C., & Winhall, J. (2020). *Building better systems: A green paper on system innovation*. Copenhagen: Rockwool Foundation, 8. https://www.systeminnovation.org/article-building-better-systems
- Loorbach, D., Frantzeskaki, N., & Avelino, F. (2017). Sustainability transitions research: transforming science and practice for societal change. *Annual Review of Environment and Resources, 42*, 599-626.
- Mosely, G., Markauskaite, L., & Wrigley, C. (2021). Design facilitation: A critical review of conceptualisations and constructs. *Thinking Skills and Creativity*, *42*, 100962.
- Overdiek, A., & Geerts, H. (2021). *Innovating with Labs: That's How You Do It! Insights from Future-Proof Retail*. The Hague: The Hague University of Applied Sciences.
- Overdiek, A., & Geerts, H. (2023). *Innovating with Labs 2.0. Creating Space for Sustainability Transitions*. The Hague: The Hague University of Applied Sciences.

- Overdiek, A. (2024). Citylab X: Towards a lens on the urban Living Lab as driver of systemic innovation. *Contexts—The Systemic Design Journal*, 2. https://doi.org/10.58279/v2002
- Schouwenberg, L. & Kaethler, M. (2021). *The Auto-Ethnographic Turn in Design.* Amsterdam: Valiz.
- Schuurman, D., De Marez, L., & Ballon, P (2016). The impact of Living Lab methodology on open innovation contributions and outcomes. *Technology Innovation Management Review* (6): 7–16.
- Stappers, P. J., & Giaccardi, E. (2017). Research through design. In *The Encyclopedia of Human-Computer Interaction* (pp. 1-94). The Interaction Design Foundation.
- Tromp, N., & Vial, S. (2023). Five components of social design: A unified framework to support research and practice. *The Design Journal, 26*(2), 210-228.
- Van Buuren, A., & Loorbach, D. (2009). Policy innovation in isolation? *Public Management Review, 11*(3), 375–392. https://doi.org/10.1080/14719030902798289
- Von Wirth, T., Fuenfschilling, L., Frantzeskaki, N., & Coenen, L. (2019). Impacts of urban Living Labs on sustainability transitions: Mechanisms and strategies for systemic change through experimentation. *European Planning Studies, 27*(2): 229-257. https://doi.org/10.1080/09654313.2018.1504895
- Williams, S., & Robinson, J. (2020). Measuring sustainability: An evaluation framework for sustainability transition experiments. *Environmental Science & Policy*, (103): 58-66. https://doi.org/10.1016/j.envsci.2019.10.012
- Yasuoka, M., Akasaka, F., Kimura, A., & Ihara, M. (2018). *Living Labs as a methodology for service design. An analysis based on cases and discussions from a systems approach viewpoint*. In DS 92: Proceedings of the DESIGN 2018 15th International Design Conference (pp. 127-136). https://doi.org/10.21278/idc.2018.0350

Aranka M. Dijkstra, Peter Joore,
Sybrith M. Tiekstra, Marije Boonstra

5. Exploring the Potential of Festivals as Living Labs for Systemic Innovation

Insights from the Interdisciplinary Innovation Program DORP

DOI: 10.1201/9781003491484

Abstract

Living Labs offer a promising approach to develop and test sustainable system innovations. One particular type of Living Lab that has received limited attention is the Festival Living Lab (FLL). Festivals can be considered as temporary mini-societies, facing systemic sustainability challenges in areas such as water, energy, housing, logistics, waste management, food, and behaviour. The temporary nature of festivals allows for adjustments to the overall societal system, allowing for experimentation with the mutual interrelationships between different aspects of the system. This makes festivals a distinctive setting for exploring sustainable system innovations. To assess the potential of FLLs as effective real-life experimentation environments, we introduce the Living Lab Activity Framework (LLAF), distinguishing various innovation stages and system levels. We utilise the LLAF to evaluate a selection of innovation projects from the DORP FLL held at the Welcome to The Village (WTTV) festival in Leeuwarden, the Netherlands. The analysis reveals that participating projects adapted their innovations based on new insights gained during various editions of the DORP FLL, demonstrating that festivals can support various stages of the innovation process on different system levels.

Introduction

Transition, system innovation, and Living Labs

Societal transitions, for instance the transition from a linear to a circular economy, are challenging processes that require innovative system changes to achieve the significant transformations in the way societal functions such as transportation, communication, housing, and feeding are fulfilled (Elzen et al., 2004). Many entrepreneurs, policymakers, and students propose innovative ideas to create a circular economy. However, only a few of these ideas are put into practice and scaled up (Kirchherr et al., 2018).

In the theory of strategic niche management (SNM), Schot & Geels (2008, p. 539) argue that many sustainable innovations face difficulties in finding market niches and user demand because they significantly differ from the prevailing technologies. To achieve a successful transition, innovation must occur at multiple system levels, fostering co-evolution and mutual adaptation between these levels (Walker & Shove, 2007). One promising approach for experimenting with such transformations in real-life settings is through Living Labs.

Living Labs can be defined as 'physical or virtual spaces where stakeholders from various sectors, including companies, public agencies, universities, and users, collaborate to create, prototype, validate, and test new technologies, services, products, and systems in real-life situations' (Leminen et al.,2012, p. 7). These Living Labs may provide an excellent environment for real-life experimentation, supporting the transition to a sustainable future.

Within the field of Living Labs, there are various sub-categories, such as Sustainable Living Labs, Product Living Labs, and Urban Living Labs (Bulkeley et al., 2016; Schliwa, 2013; Steen & van Bueren, 2017). Another rather innovative category is Festival Living Labs (Boonstra et al., 2021). This study explores the potential of this specific type of labs to facilitate sustainable system innovation, as previously has been presented (Dijkstra et al., 2023).

Festival Living Labs

While there is no extensive scientific literature specifically focussed on Festival Living Labs (FLLs), there are ongoing initiatives exploring innovation at festivals. Several European festivals allow scientific research and testing of new innovations on their premises (de Ruiter, 2012; Open-House, 2019; Stichting Innofest, 2019). Additionally, several regional and European-funded projects are investigating the concept of festivals as test beds for innovation (Stichting Innofest, 2019; Inno-Quarter, 2019). These initiatives highlight the unique characteristics of festivals that make them ideal for experimentation, which have for instance been described in the Festival Experimentation Guide (Dijkstra & Boonstra, 2021).

Festivals may be defined as celebratory events that build one or more temporary, independent logistical infrastructures, such as an energy grid, a camp site, and/or water supply for the purpose of facilitating the gathering of people. Combining the definition of a festival with the definition of a Living Lab, a Festival Living Lab may be defined as a celebratory event that facilitates the gathering of people and that offers (interdisciplinary collaborations between) companies, public agencies, universities, users, and other stakeholders access to one or more of their temporary, independent logistical infrastructures to create, prototype, validate, and test new technologies, services, products, and systems.

Festival sites resemble small cities or temporary settlements, with temporary inhabitants engaging in activities like eating, sleeping, and waste generation. These temporary 'mini societies' face similar sustainability challenges related to water, energy, housing, logistics, waste management, food, and behaviour. For example, the three-day LowLands festival uses 300.000 kWh of electricity generated with 120.000 litres of Diesel each edition (LowLands, 2019). The unique aspect of festival sites, compared to cities, is their temporary and flexible nature. Since festivals are built from scratch for each event, it is relatively easy to make adjustments and interventions to experiment with different systems. The clearly defined borders of the festival site allow for better control and monitoring of material and energy flows. Additionally, festivals provide an affordable platform for experimentation, particularly for smaller companies focused on impact-oriented projects (Elks, 2019).

Festivals offer a unique opportunity to address behavioural change and social acceptance, which are essential for the transition to a circular economy. While attending festivals, people are expected to be more open to trying new things, making festivals ideal for introducing novelties (Potts, 2011; Schulte-Römer, 2013). From a Living Lab perspective, festivals attract a large audience that can be engaged as end-users in open innovation processes and experiments (Leminen, 2015). Moreover, the repetitive nature of some festivals may create an opportunity for a consecutive chain of FLLs, enabling multiple iterative experiments within slightly different settings in a relatively short period of time.

Evaluating system innovation at Festival Living Labs

Festivals have characteristics that make them ideal for experimentation and promoting sustainability transitions. With several FLLs already in operation, this offers an opportunity to examine the effectiveness and impact of these FLLs on sustainable system innovation. Therefore, this study aims to answer the following research question:

Are Festival Living Labs effective real-life experimentation settings for sustainable system innovation?

Evaluating FLLs poses a challenge. Several studies emphasise the need for standardised methods to assess the effectiveness, impacts, and performance of Living Labs (Beaudoin et al., 2022). A scoping review by Bronson et al. (2021) reveals that there is currently no widely applicable approach or framework for evaluating the impact of Living Labs. The dominant method described in literature is comparative qualitative case studies, assessing the overall functioning and achievement of specific Living Labs. However, since our focus is on evaluating the impact of experiments within FLLs, none of the existing frameworks adequately address our research question. We therefore introduce the Living Lab Activity Framework (LAFF) as an evaluative framework to assess the DORP Festival Living Lab. By discussing the results, we provide insights into the research question and conclude with recommendations for future studies. Through this work, we aim to contribute to the existing knowledge on evaluating and assessing the outcomes and impact of Living Labs.

Method

To explore the potential of Festival Living Labs for system innovation, we employed various research methods. Our approach can be summarised in the following steps:

A. Develop Evaluative Framework: We conducted a literature review to identify existing evaluative approaches and frameworks for Living Labs. Based on this review, we developed the Living Lab Activity Framework (LLAF) to evaluate the progress of innovation projects in a Festival Living Lab across different stages and levels of the system.
B. Case Selection: We chose the DORP Festival Living Lab as a case study to apply the LLAF. This selection was based on the authors' close involvement in this specific Festival Living Lab and access to relevant documentation. We used existing project documentation to describe the DORP Festival Living Lab.

C. Project Selection: Over the years, numerous projects participated in the DORP Living Lab. To determine which projects to include in the LLAF analysis, we compiled a comprehensive inventory of projects that took place within the DORP Living Lab from 2015 to 2018. We then applied the following criteria to select projects for analysis: (a) Projects that occurred between 2016 and 2018 were included, as projects from the first pilot year of the DORP Summer School deemed not representative when the DORP program was still developing itself as a FLL, (b) Student projects were excluded, as they were primarily educational experiences with limited follow-up, and (c) Projects with insufficient or incomplete data were also excluded. This resulted in 31 projects to be analysed.
D. Plotting and Analysis of Projects: The selected projects were mapped onto the LLAF by determining their innovation stage and system level before and after their participation in the DORP Living Lab. The categorization of projects into different innovation stages and system levels was done through an iterative process involving the researchers, who observed these projects in the field. The first author was involved as program leader of DORP. We used predefined criteria described in Tables 3 and 4. The resulting mapping is presented in Figure 1. Subsequently, the authors evaluated the activities of the projects within the DORP Living Lab through discussion, interpreting how and if the projects progressed between innovation stages and system levels.

Living Lab Activity Framework

In the introduction, we argue that transitioning to sustainable systems requires radical innovation, which often happens in niche areas where system levels mutually adapt (Sengers et al., 2019). Living Labs are seen as a way to experiment with these adaptations. Therefore, projects in Living Labs should focus on (a) experimentation and (b) exploring interdependencies between system levels. To evaluate the potential of using festivals as real-life settings for sustainable system innovation, we have identified two sub-questions for our evaluation framework:

- *Sub-question I: What phase of the innovation process do projects in the Festival Living Lab focus on?*
- *Sub-question II: What system level do projects in the Festival Living Lab focus on?*

We acknowledge that design and innovation processes are non-linear, lacking a predetermined sequence for addressing specific system levels within innovation. While there's a resemblance between the hierarchical approach of system levels and the means-end chain often used by designers (Joore & Brezet, 2015), innovation processes, influenced by factors like technology, legislation, and user markets, develop interdependently. This holistic approach is evident in models such as Acklin's Design-Driven Innovation Process (2010) and the TU/e Innovation Lab model (Den Ouden et al., 2016). Real-life testing in Living Labs unveils these interdependencies, allowing projects to start with a focus on a specific aspect or system level but adapt as needed based on feedback and challenges from other levels or aspects. The iterative learning process involves conducting experiments, monitoring outcomes, and making improvements, generating valuable knowledge in a real-life setting (Schliwa, 2013). The ability to iterate between innovation stages and system levels adds value to the design process in Living Labs.

To show the iterative nature between innovation stages and system levels, we created a matrix, forming the Living Lab Activity Framework (LLAF) as depicted in Figure 1. In this framework, both sub-questions correspond to the two axes: system levels (sub-question I) are represented along the y-axis, and the project's innovation stages (sub-question II) are represented along the x-axis. Based on literature (refer to Tables 1 and 2), we identified five innovation stages on the x-axis (1. Exploration, 2. Development, 3. Experimentation, 4. Implementation, and 5. Commercialisation) and four system levels on the y-axis (A. Product-Technology System, B. Product-Service System, C. Socio-Technical System, and D. Societal System).

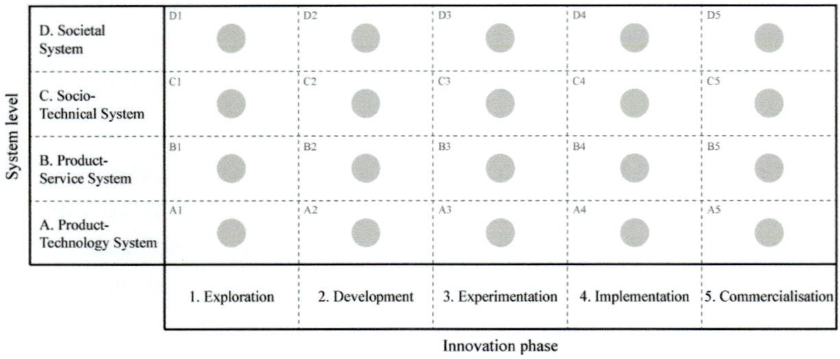

Figure 1. Living Lab Activity Framework (LAFF) for evaluating Festival Living Labs projects (Dijkstra et al., 2023).

The LAFF is used to plot a project's innovation phase and system level on the framework before and after participating in the Festival Living Lab, using the criteria described in Tables 3 and 4. This provides a visual representation of a project's development in the Living Lab.

Although Living Labs involve diverse actors, resources, and activities that support innovation throughout the lifecycle (Leminen et al., 2012), projects in Living Labs can only undergo a limited number of iterations within their programs and settings. Therefore, the framework only shows the progress of the research, development, and/or experimentation process made by projects within one or more editions of the Festival Living Lab itself. It does not indicate the overall impact of the Living Lab on a project's innovation progress. Additionally, since different programs within our case study focus on different types of challenges (e.g., technical prototypes, business models, user behaviour), the phase a project starts or ends in does not reflect the quality of iterations made by the projects. Therefore, the iterations made by projects are not qualitatively comparable.

Table 1. Comparison of innovation stages in literature (x-axis) (Dijkstra et al., 2023).

<table>
<tr><th>Living Lab Activity Framework</th><th>TRLs (Mankins, 1995)</th><th>Stage-Gate Model (Cooper et al., 2002)</th><th>The Fugle Model's Innovation Funnel (Du et al., 2008)</th><th>TUe Innovation Lab (den Ouden, 2016)</th><th>ULL Way of Working (Steen & van Bueren, 2017)</th></tr>
<tr><td rowspan="2">1. Exploration</td><td>TRL0 – Not officially defined by NASA</td><td>1. Preliminary assessment</td><td>A. Idea Generation/ Identification</td><td rowspan="2">Exploration</td><td rowspan="2">Research</td></tr>
<tr><td>TRL1 – Basic Research</td><td>2. Definition</td><td>B. Concept Definition</td></tr>
<tr><td rowspan="2">2. Development</td><td>TRL2 – Proof of Principle</td><td rowspan="3">3. Development</td><td>C. Concept Feasibility and Refinement</td><td rowspan="2">Concept Development</td><td rowspan="2">Development</td></tr>
<tr><td>TRL3 – Early lab scale demon-stration</td><td>D. Portfolio</td></tr>
<tr><td rowspan="2">3. Experi-mentation</td><td>TRL4 – Lab scale demon-stration</td><td rowspan="2">E. Deployment</td><td rowspan="3">Evaluation and Validation</td><td rowspan="2">Testing</td></tr>
<tr><td>TRL5 – Validation</td><td rowspan="4">4. Validation</td></tr>
<tr><td rowspan="2">4. Implemen-tation</td><td>TRL6 – Early prototype</td><td rowspan="3">F. Refinement and Formalisation</td><td rowspan="2">Imple-mentation</td></tr>
<tr><td>TRL7 – Late prototype</td><td rowspan="3">Market Introduction</td></tr>
<tr><td rowspan="2">5. Commercial-isation</td><td>TRL8 – Early stage commercial environment application</td><td rowspan="2">Commercial-isation</td></tr>
<tr><td>TRL9 – Market ready application full commercial application</td><td>5. Commercial-isation</td><td>G. Exploitation Stage</td></tr>
</table>

Table 2. Comparison of system levels in literature (y-axis) (Dijkstra et al., 2023).

Living Lab Activity Framework	Innovation levels (Ceschin & Gaziulusoy, 2016)	MDM Model (Joore & Brezet, 2015)	Transition Management (Geels, 2005)	Intelligent Products (Andrews, 2003)	Design for Sustainability (Brezet et al., 2001)	Systems Engineering (Haugan, 2001)	Means-end-chain (Roozenburg & Eekels, 1998)
D. Societal System	Spatio-Social innovation level	S: Societal System	Transitions (landscape)	Rethinking Values	System innovation	System	Values
C. Socio-Technical System	Socio-Technical System innovation level	R: Socio-Technical System	System innovation (social-technical regime)	Systemic Context	Function innovation	Sub-system	Needs
B. Product-Service System	Product-Service System innovation level	Q: Product-Service System	Process innovation (niche)	Ecological Context	Function Redesign	Element	Functions characteristics
A. Product-Technology System	Product innovation level	P: Product-Technology System	Product innovation (niche)	Immediate Context	Product Improvement	Component	Form

Table 3: LLAF innovation stages (x-axis) (Dijkstra et al., 2023).

Dimension	Description	Criteria
1. Exploration	The process of making new discoveries about a problem or solution and coming up with an innovative concept.	The project is based on an idea or problem but has no evidence to base its assumptions on. It is an unproven concept and no validation has been done yet.
2. Development	The process of advancing basic ideas and concepts into more concrete and holistic requirements of the innovation.	The project is based on a clear concept but needs further development and/or validation of its underlying assumptions.
3. Experimentation	The process of testing and validating assumptions about the innovation.	The project has a prototype that needs to be tested. This can be a physical prototype but also, e.g., a service or societal concept.
4. Implementation	The process of applying or integrating the innovation in its designated real-life setting.	The project has a product, service, or approach that is tested in relation to its context while being integrated in the larger system.
5. Commercialisation	The process of making the innovation available on the market.	The project has a product, service, or approach that is implemented and commercially operates in its (simulated) context, testing mutual dependencies between all system aspects (technical, economic, and social).

Table 4. LLAF system levels (y-axis) (Dijkstra et al., 2023).

Dimension	Description	Festival Context	Criteria
A. Product-Technology System	The Product-Technology System level is made up of tangible products that one can touch.	Within the context of a festival, the product level refers to the 'hardware' the festival is built up from (tents, cabins, sound systems, generators, etc).	The project focuses on tangible products.
B. Product-Service System	The Product-Service System level is made out of the combination of physical and organisational components that together fulfil a specific function.	Within the festival, the service level refers to the services provided for by the festival; the total of products and services providing, e.g., the economic infrastructure (often coins) people can buy food or drinks with, the campsite people can safely sleep in, but also the provision of drinking water and the service of waste removal.	The project focuses on new types of services (e.g., Product-as-a-Service models, cryptocurrency systems) and/or exploring their product's market fit.
C. Socio-Technical System	At the Socio-Technical System level 'a large number of components are combined that are not formally related to each other' (Joore & Brezet, 2015). The socio-technical system can be defined as 'a cluster of aligned elements, including artefacts, technology, knowledge, user practices and markets, regulation, cultural meaning, infrastructure, maintenance networks and supply networks, that together fulfil a specific societal function' (Geels, 2005).	Within the context of a festival, this level refers to the coherence of the festival's technical and economic infrastructure together with its entertainment program, its safety protocols, its organisation, suppliers and stakeholders, and its audience.	The project focuses on the integration of new products or services in (a part of) the full festival system. An important difference within this criterion as opposed to experimentation on other system levels, is that something in the wider system of the festival is depending on the project innovation's functioning.
D. Societal System	The Societal System level relates to the intangible believes, traditions, norms and values of a community of people in a specific place.	Within the festival context the Societal System level is made up of the festival audience that behaves according to its communal believes.	The project focuses on behavioural change or the acceptance of change.

Results

Description of the DORP Festival Living Lab case

The DORP Festival Living Lab was a part of Welcome to The Village (WTTV), a popular music festival that took place over three days in Leeuwarden, a city in the north of the Netherlands. The festival attracted around 9,000 visitors who enjoy music performances on three stages, along with theatre shows, visual arts, and a program focused on social issues, sustainability, and innovation. From 2014 to 2018, more than 70 innovation projects by students, start-ups, and companies were featured in various innovation programs associated with the WTTV festival. These programs aimed to develop, test, and promote sustainable concepts, prototypes, business models, and service models. Each program focused on projects at different stages of the innovation process and utilised the festival in different ways to support and accelerate sustainable innovations. Together, we refer to all these programs as the DORP Festival Living Lab (DORP meaning VILLAGE in Dutch).

Test and implementation projects

In the DORP Festival Living Lab, entrepreneurs had the opportunity to test their innovations using the festival's infrastructure and audience. Innofest, a platform supporting innovation at festivals, facilitated several projects that used the festival as a real-life testing ground. For example, **Greener** tested their off-grid battery, which provides sustainable energy for festivals as an alternative to diesel generators commonly used. Another project, **Loyal Garden**, developed a blockchain system to pay festival volunteers with a specific cryptocurrency. A prototype of the system was implemented during the DORP Summer School prior to the festival, where participants used a personal QR code on their festival wristbands to purchase drinks at the DORP bar. During the festival, the system was tested with volunteers in the backstage area.

Additionally, as the festival organiser, WTTV sometimes acts as an early adopter of new sustainable and circular innovations, particularly those related to festivals. One example is the collaboration with **LILY**, a light installation initially designed to illuminate the pathway from the festival to its campsite. Over the years, LILY has been extensively tested and further developed at the WTTV festival. It has evolved into a floating art installation inspired by natural patterns such as schools of fish or flocks of birds. Illustrations of these projects can be seen in Figure 2.

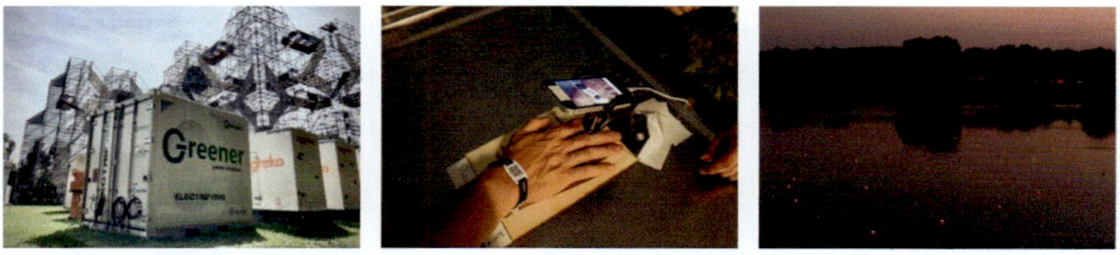

Figure 2. Illustrations of test and implementation projects at WTTV: Greener (Picture © Greener), Loyal Garden (Picture © Innofest), and LILY (Picture © WERC).

Innovation projects

Within the DORP Festival Living Lab, there was a sub-program called the DORP Summer School. This program used the festival as a place for collaboration and co-creation. The DORP Summer School was created to give entrepreneurs and organisations the chance to further develop their innovative ideas and concepts with the help of a diverse team of students from different disciplines, faculties, and universities. The students work together with the entrepreneurs or organisations for 7 days, guided by experts, using a hackathon format based on the design thinking approach.

During the Summer School, the interdisciplinary teams assist in developing and validating the concepts or prototypes directly at the festival. This quick feedback loop sets it apart from other hackathon programs that usually focus on either idea generation or development phases. From a university perspective, the Summer School serves as an interdisciplinary course that teaches students how to work together effectively in interdisciplinary teams.

The challenges brought to the DORP Summer School can cover various aspects and may be in different stages of innovation. For example, there may be a need for scientific research to develop innovative concepts, such as the project **Offgrid Basecamps** brought in by construction company Van Wijnen. In this challenge, the team worked on finding a solution for construction site managers to select the best renewable energy option for their sites.

On the other hand, entrepreneurs may already have a technical prototype that needs further development and testing. For instance, **Saru Soda** sought assistance in modifying a post-mix lemonade machine to dispense their organic lemonades. Another example is **Comp-A-Tent**, who aimed to create an appealing and functional festival tent using their newly patented compostable material. You can see illustrations of these projects in Figure 3.

Figure 3. Illustrations of innovation projects at WTTV: Saru Soda (Picture © Nena Bode), Comp-A-Tent (Picture © DORP Summer School), and Offgrid Basecamps (Picture © DORP Summer School).

Experience projects

Festivals provide an exciting opportunity to introduce new and innovative ideas. To involve festival-goers in the innovation process and gather feedback for the projects, a designated area within the WTTV festival site is dedicated to innovation. This space serves as a platform to showcase new products, business models, and services in a fun, interactive, and accessible manner. Its aim is to raise awareness and generate support for sustainable changes. Here, festival attendees become unwitting participants in scientific research or provide feedback on the new offerings from entrepreneurs.

One example of such a project is the **Hair-Washing District** created by the Japanese artist Sachi Miyachi. In collaboration with students from the DORP Summer School, they designed a raised and self-sustaining structure where festival-goers could have their hair washed, encouraging them to appreciate life's simple pleasures. Another instance is the **Snackathon** introduced by WTTV in 2018. In the Snackathon, food entrepreneurs were challenged to develop healthy and sustainable snacks for the 'Cafeteria of the Future' during the DORP Summer School. These snacks were then sold directly to festival attendees, allowing them to sample and provide feedback. Examples of the snacks created include 'cricket fries,' made from cricket flour by **&Cricket**, and the **Vegandel**, a vegan twist on a traditional Dutch snack using seitan. You can see illustrations of these projects in Figure 4.

Figure 4. Illustrations of Experience projects at WTTV: Hair-Washing District (Picture © Nena Bode), Vegandel (Picture © DORP Summer School), and &Cricket (Picture © DORP Summer School).

Resulting plotting of DORP projects

Plotting the selected DORP projects on the Living Lab Activity Framework (LAFF) helps visualise their progress. You can see the results in Figure 5.

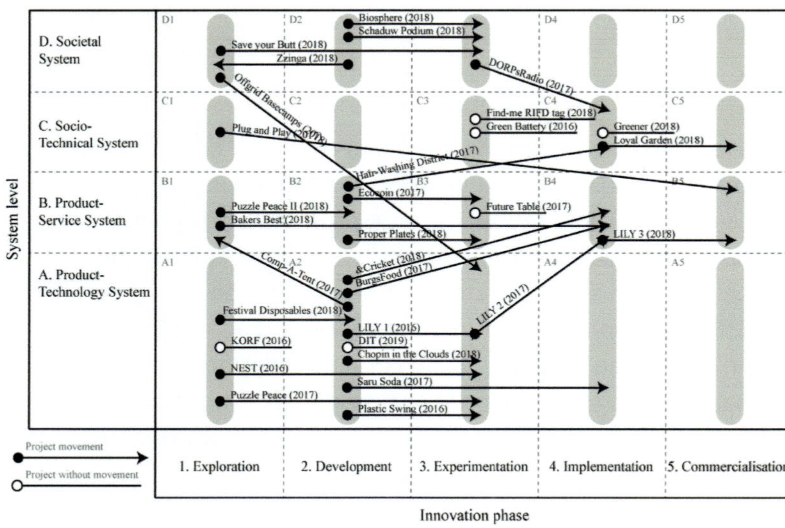

Figure 5. Selected projects participating in the DORP Living Lab from 2016 to 2018 plotted on the Living Lab Activity Framework (Dijkstra et al., 2023).

Figure 5 shows that many projects have made significant progress in their innovation process by testing, implementing, or even commercialising their products, services, or concepts at the festival (16 out of 31 projects). For example, **Saru Soda** moved from Product-Technology Development (A2) to Product-Technology Implementation (A4).

The framework also highlights that some projects iterate between different levels (7 out of 31 projects). On one hand, there are forward iterations. For instance, **Offgrid Basecamps** started with research on selecting the best renewable energy solution for construction sites. They developed a decision-based algorithm (D1) and then created a prototype of a serious game, which was tested at the festival (A3) among the audience. Another project with significant iteration is **Plug & Play**, which transitioned from the Exploration phase on the Socio-Technical level (C1) to the Commercialisation phase on the Product-Service level (B5). Plug & Play aimed to explore how electric car batteries could power music stages in the future. At the festival, the students working on this challenge successfully 'hacked' an electric car and organised a fully operational car-powered silent disco. Projects that also made quite large iterations are **BurgsFood** and **&Cricket**. These two projects were part of the Snackathon and developed, tested, and eventually sold sustainable snacks at the festival. As they were challenged to sell their snacks according to the official festival rules (so they would not be unfair competition to other food stalls) these projects were really forced to make a leap from Development on a Product-Technology level (A2) to Implementation on the Product-Service level (B4).

On the other hand, some projects iterate backward in the framework (2 out of 31 projects), namely **Zzinga** and **Comp-A-Tent**. This backward movement does not mean that no valuable insights were found, but rather that they encountered challenges during the program that required them to reassess the viability of their ideas in their current form. This was one of the objectives of the DORP Summer School: to identify early on whether an innovative concept is feasible before investing significant time and resources into its development. For example, **Comp-A-Tent** aimed to design and test a new biodegradable tent for festival-goers based on their patented material. However, during the design process, they discovered that their intended user (the festival-goer) was not their actual customer. The festival organisation itself became their customer, leading to changes in the requirements and the overall business case. Despite this shift, the DORP Living Lab still provided valuable insights for Comp-A-Tent, albeit on a different system level than initially intended.

Not directly visible but worth noting is that projects that participated in the DORP Living Lab for multiple years focused on challenges at different system levels each year. A prime example is the **LILY** project by WERC, which was present at the WTTV festival every year. It evolved from a single LILY prototype in 2016 to a fully implemented sustainable art installation in 2018. Another example is **Puzzle Peace**, which initially joined the DORP Living Lab in 2017 with a challenge to develop multifunctional furniture. They successfully created a prototype that the festival organisation purchased as a launching customer. The following year, they returned to the DORP Summer School to develop their business case, which they then tested among the festival audience.

Discussion

The aim of this study is to understand whether FLLs may function as effective real-life experimentation settings for sustainable system innovation. Our research resulted in three key findings:

1. Festival Living Labs may function as a relevant real-life experimentation setting for sustainable innovation.

Festival Living Labs can serve as practical settings for testing sustainable innovations. The proposed LLAF supported visualising the progress of innovation projects, showing how they move horizontally between different innovation stages and vertically across system levels. The analysis of 31 DORP projects indicates that Festival Living Labs, like DORP, can support learning across innovation phases and navigate between system levels. This is crucial because achieving radical change for sustainable transition requires a holistic view and coordination across system levels (Walker & Shove, 2007; Schot & Geels, 2008). The DORP Festival Living Lab is a unique initiative linked to the identity of the WTTV festival. To better understand whether all festivals can support sustainable system innovation, further research on the characteristics and requirements of Festival Living Labs and other types of Living Labs is necessary.

2. Festival Living Labs (FLLs) have the potential for system innovation, but this has not been conclusively proven.

The dynamic and adaptable nature of festivals makes them ideal for experimenting with technical, economic, and social systems (Dijkstra & Boonstra, 2021). In the LLAF, projects exploring these systems corresponds to sections C3 and C4. However, in the DORP FLL, only a few projects fall into this category. Analysis indicates that most development in DORP FLL focuses on product-technology systems (level A) and product-service systems (level B).

This aligns with findings by Steen & Van Bueren (2017) that many Urban Living Labs lack key characteristics for groundbreaking innovations. The LLAF also shows some experimentation with the Societal System (level D), supporting the idea that festivals are suitable for trying novel ideas (Potts, 2011; Schulte-Römer, 2013).

The emphasis on the lower left corner of the LLAF may suggest FLLs are not effective for sustainable system innovation. However, the limited projects in this area could be influenced by research limitations, such as incomplete data plotting and different perspectives from project owners. DORP FLL programs primarily focus on accelerating innovation in general, not specifically on Socio-Technical System innovation. Thus, while our study indicates FLLs can be effective for sustainable innovation projects, it doesn't necessarily confirm festivals as particularly effective for sustainable system innovation.

To better understand FLLs' effectiveness for sustainable system innovation, case studies specifically focusing on Socio-Technical System innovation are needed. These studies would explore how movements on the LLAF towards and from the Socio-Technical system level can be facilitated.

3. The LLAF may contribute to analysing the effectiveness of Living Labs by providing a framework to evaluate and compare their impact over time.

The LLAF can help assess the effectiveness of Living Labs by providing a framework to evaluate and compare their impact over time. There is a growing need for methods and frameworks to evaluate Living Labs' impact and effectiveness (Beaudoin et al., 2022; Bronson et al., 2021). Using the LLAF to visualise the progress of Living Lab projects may offer deeper insights into their outcomes, allowing for potential improvements in focus or design to enhance innovation results. However, it's important to note that the LLAF only captures project iterations within one or more editions of a Festival Living Lab (FLL). It doesn't reveal whether a project's involvement in the FLL has a lasting impact on its long-term innovation process outside of the FLL event. Assessing the FLL's impact beyond the event is beyond the scope of this chapter. To validate the effectiveness and applicability of the LLAF, further research should apply the framework to other FLLs or Living Labs in different contexts (e.g., various Urban Living Labs). Additionally, exploring how FLL experiment results and insights may be scaled beyond the FLL would be valuable.

Conclusion

The purpose of this study was to explore whether Festival Living Labs (FLLs) can be effective settings for system innovation. We wanted to investigate if festivals, with their infrastructure, can support real-life experiments on the Socio-Technical System level and contribute to accelerating sustainable transitions. To answer this question, we developed the Living Lab Activity Framework (LLAF) to help identify the innovation stage and system level of projects participating in an FLL. By analysing 31 innovation projects from the DORP Living Lab at the Welcome to The Village festival in the Netherlands, we found that projects were present at all system levels. This suggests that FLLs can indeed facilitate experiments across different levels. However, most of the projects in the DORP Living Lab primarily focused on developing products and services. Therefore, the findings don't strongly support the idea that festivals are specifically effective for sustainable system innovation. Nevertheless, the study indicates that festivals can be suitable environments to function as Living Labs for various types of innovation, including system innovation. Future research could explore how movements towards and from the Socio-Technical system level can be facilitated. To gain a deeper understanding of whether festivals are suitable contexts to support sustainable system innovation, more research on FLLs and other types of Living Labs is necessary, with a specific focus on system innovation and sustainable transitions.

Bibliography

- Acklin, C. (2010). Design-Driven Innovation Process Model. *Design Management Journal, 5*(1), 50–60. https://doi.org/10.1111/j.1948-7177.2010.00013.x
- Andrews, A. (2003). Networks, systems and society. In E. Aarts & S. Marzano (Eds.), *The New Everyday: Views on Ambient Intelligence*. Koninklijke Philips NV, Eindhoven.

- Beaudoin, C., Joncoux, S., Jasmin, JF., Berberi, A., McPhee, C., Schillo, R.S. & Nguyen, V.M. (2022). A research agenda for evaluating Living Labs as an open innovation model for environmental and agricultural sustainability. *Environmental Challenges, 7*, 100505. ISSN 2667-0100.
- Boonstra, M., Dijkstra, A., & Joore, P. (2021). Festival Living Labs: Involving the Festival Community in Sustainable Experimentation. In *DESIGN CULTURE(S) Cumulus Conference Proceedings Roma 2021 Volume #2* (page 4783).
- Brezet, J., Bijma, A., Ehrenfeld, J., & Silvester, S. (2001). The Design of Eco-efficient Services. Method, Tools and Review of the Case Study Based 'Designing Eco-efficient Services' Project. Ministry of VROM and Delft University of Technology, NL.
- Bronson, K., Devkota, R., & Nguyen, V. (2021). Moving toward Generalizability? A scoping review on measuring the impact of Living Labs. *Sustainability, 13*(2), 502. doi: 10.3390/su13020502.
- Bulkeley, H., Mai, L., Coenen, L., Frantzeskaki, N., van Steenbergen, F., Hartmann, C., ... Voytenko Palgan, Y. (2016). Urban Living Labs: Governing urban sustainability transitions. *Current Opinion in Environmental Sustainability, 22*, 13–17. https://doi.org/10.1016/j.cosust.2017.02.003
- Ceschin, F., & Gaziulusoy, A. I. (2016). Evolution of design for sustainability: From product design to design for system innovations and transitions. *Design Studies, 47*, 118-163. https://doi.org/10.1016/j.destud.2016.09.002
- Dijkstra, A.M., & Boonstra, M. (2021). *Festival Experimentation Guide* (1st ed.). Leeuwarden, NL: NHL Stenden University of Applied Sciences (ISBN: 9789491589973).
- Dijkstra, A. M., Tiekstra, S. M., Boonstra, M., & Joore, P. (2023). Festivals as Living Labs for System Innovation: Experiences from the interdisciplinary innovation programme DORP. In D. Schuurman (Ed.), *Proceedings of the OpenLivingLab Days Conference 2023: 'Living Labs for an Era of Transitions' How human-centric innovation is changing our lives* (pp. 28-47). European Network of Living Labs.
- Cooper, B. R. G., Edgett, S. J., & Kleinschmidt, E. J. (2002). Optimizing the Stage-Gate Process: What Best Practice Companies are Doing – Part Two. *Stage-Gate Inc, 45*(5).

- De Ruiter, J. (2012). Duurzame activiteiten op Llowlab trekken zeker 13.000 nieuwsgierige bezoekers tijdens Lowlands festival. Retrieved July 27, 2019, from Double2 BV website: http://d2bv.nl/site/duurzame-activiteiten-op-llowlab-trekken-zeker-13-000-nieuwsgierige-bezoekers-tijdens-lowlands-festival/
- Den Ouden, E., Valkenburg, R., & Blok, S. (2016). Exploring the Future of Living Labs. In LightHouse, TU/e Innovation Lab. https://doi.org/10.1080/16549716.2017.1397909
- Du Preez, N. D., & Louw, L. (2008). A framework for managing the innovation process. In *PICMET: Portland International Center for Management of Engineering and Technology, Proceedings*, (pp. 546–558). https://doi.org/10.1109/PICMET.2008.4599663
- Elks, S. (2019, June 25). Festival or refugee camp? Music events test emergency aid. *Thomson Reuters Foundation*. Retrieved from https://www.reuters.com/article/us-global-climate-festivals-feature/festival-or-refugee-camp-music-events-test-emergency-aid-idUSKCN1TR24K
- Elzen, B., Geels, F. W., & Green, K. (2004). System innovation and the transition to sustainability: theory, evidence and policy. *Edward Elgar Publishing*.
- Geels, F. W. (2005). Processes and patterns in transitions and system innovations: Refining the co-evolutionary multi-level perspective. *Technological Forecasting and Social Change, 72*(6 SPEC. ISS.), 681–696. https://doi.org/10.1016/j.techfore.2004.08.014
- Haugan, G. (2001). Effective Work Breakdown Structures. *Management Concepts Inc, Vienna, VA, USA*.
- Inno-Quarter. (2019). What Is Inno-Quarter? Retrieved July 28, 2019, from https://www.innoquarter.eu/about/about/
- Joore, P., & Brezet, H. (2015). A Multilevel Design Model: The mutual relationship between product-service system development and societal change processes. *Journal of Cleaner Production, 97*(March), 92–105. https://doi.org/10.1016/j.jclepro.2014.06.043
- Kirchherr, J., Piscicelli, L., Bour, R., Kostense-Smit, E., Muller, J., Huibrechtse-Truijens, A., & Hekkert, M. (2018). Barriers to the Circular Economy: Evidence From the European Union (EU). *Ecological Economics, 150*(August), 264–272. https://doi.org/10.1016/j.ecolecon.2018.04.028
- Leminen, S. (2015). Living Labs as Open Innovation Networks. https://doi.org/10.13140/RG.2.1.2423.5281

- Leminen, S., Westerlund, M., & Nyström, A.-G. (2012). Technology Innovation Management Review Living Labs as Open-Innovation Networks. *Technology Innovation Management Review, September*(September), 6–11. Retrieved from www.timreview.ca
- LowLands. (2019). LowLands Green & Clean. Retrieved July 28, 2019, from LowLands Festival website: https://lowlands.nl/green-clean/
- Mankins, J. C. (1995). Technology readiness levels. *White Paper, April, 6* (1995).
- Open-House. (2019). What Does Open-House Do? Retrieved July 28, 2019 from https://www.open-house.nl/what-we-do/
- Potts, J. (2011). *Creative Industries and Economic Evolution*. Edward Elgar Publishing.
- Roozenburg, N., & Eekels, J. (1998). *Product Design: Fundamentals and Methods* (2nd ed.). Wiley Chichester.
- Schliwa, G. (2013). Exploring Living Labs through Transition Management – Challenges and Opportunities for Sustainable Urban Transitions. *IIIEE Master Thesis, (September)*, 80. Retrieved from http://lup.lub.lu.se/student-papers/record/4091934/file/4091935.pdf
- Schot, J., & Geels, F. W. (2008). Technology Analysis & Strategic Management Strategic niche management and sustainable innovation journeys : theory , findings , research agenda. *Strategic Niche Management Research, 7325*(August), 537–554. https://doi.org/10.1080/09537320802292651
- Schulte-Römer, N. (2013). Fair framings: Arts and culture festivals as sites for technical innovation. *Mind and Society, 12*(1), 151–165. https://doi.org/10.1007/s11299-013-0114-8
- Sengers, F., Wieczorek, A. J., & Raven, R. (2019). Experimenting for sustainability transitions: A systematic literature review. *Technological Forecasting and Social Change, 145*, 153–164. https://doi.org/10.1016/j.techfore.2016.08.031
- Steen, K., & van Bueren, E. (2017). The defining characteristics of urban Living Labs. *Technology Innovation Management Review, 7*(7), 21–33. Retrieved from https://timreview.ca/article/1088
- Stichting Innofest. (2019). Hoe Werkt Het: Test Prototypes of Innovaties tijdens één van onze festivals. Retrieved July 28, 2019, from Innofest website: https://innofest.co/over/hoe-werkt-het/
- Walker, G., & Shove, E. (2007). Caution! Transitions ahead: politics, practice and sustainable transition management. *Environment and Planning A, 39*(1998), 763–770. https://doi.org/10.1068/a39310

Part 2: Social Dynamics in Living Labs

Wina Smeenk, Perica Savanović,
Marieke Zielhuis, Daan Andriessen

6. Opening & Closing Hours

Three Cases and Their Dynamics to Let Learning Thrive

DOI: 10.1201/9781003491484

Introduction

Collaboration in experimental learning, innovation, and transition environments such as Living Labs can be complicated to carry out in practice. Designers and design researchers are often involved in these processes based on their intrinsic emphasis on transformation of the status quo towards new ways of collaboration for sustainable common futures. We propose that these new ways of collaboration have to include joint learning processes. In this chapter, we reflect on the dynamics within such environments which affect the way they are open or closed in their learning processes.

We propose that Living Labs should not just be characterised as 'closed' when they only involve a limited amount of participants, such as a selection of users, or as 'open' when they involve more and diverse people, companies, etc. We propose that, seen from the perspective of learning dynamics, opening and closing entails more than just 'who is involved'. It is also about the way those involved are able to learn together.

In three different cases, we show different ways of working and learning by the involved parties when it comes to opening and closing dynamics. The dynamics between the different collaborating partners in each case are discussed using the four learning mechanisms in boundary crossing processes: identification, reflection, coordination, and transformation.

The chapter concludes that it is important that the different involved parties in a lab should together to address, decide on, and reflect on the decisions which affect the dynamics within a lab in which joint learning can thrive. We distinguish several topics for such considerations, such as goals, design space, information availability, and/or decision power. The insights from these three cases can be helpful for anyone pursuing collaborative transformations and striving to let learning thrive.

The Need for Approaches To Open up Learning

Our society is currently facing a number of grand challenges and transitions. These challenges even require the fundaments of our society to be revisited in order to keep it livable, resilient, and meaningful (e.g., Chen et al., 2016). In order to address these challenges in a timely manner, concrete actions are necessary. Within the built environment, where two of the three cases in this chapter are set, the socio-technical challenges are complex but real and visible to everybody. These include transition in materials and energy(use), mobility, safety, housing, space allocation, climate adaptation, and sustainable living spaces in general, all of which need practical and concrete actions involving all stakeholders in order for change and transition to happen.

Within the built environment, different traditional (professional) disciplines in the construction sector are rather risk-averse and reactive. This attitude is often prompted by strict divisions (in roles), starting between clients and contractors, and progressing to specialist disciplines. In addition, there is a paradox, prompted by the way we educate engineering and construction professionals. Building engineers are traditionally trained to propose solutions to well-defined technical problems. Their added value is apparent through the objective use of knowledge, so that not professionals themselves but their solutions come to the fore. As a consequence, they try to do their work from relative anonymity on the basis of agreed objectives and criteria that are usually translated in as specific as possible programs of requirements. This puts engineers and builders in a well-defined role, but one that is also difficult to change if the demand changes — as often is the case during innovation processes tackling open socio-technical challenges.

Continuing to seek refuge in the familiar rational efficiency of the traditional approach will therefore have to change; in any case more quickly than it incrementally does in the current practice. Our view is that professionals in general, and engineers specifically, should move more towards expansive co-design approaches, instead of further optimising their rational reductionistic practices.

Savanović (2021, p.169,) "argues that the built environment needs to learn how to incorporate the openness of the design process into the traditional and still prevailing engineering and construction processes. The creation, from a quadruple-helix perspective and active collaboration, of new (design) opportunities, options, and possibilities for alternative and joint sustainable future(s) together needs to precede (shared)decision making on which possibilities to further pursue or use. Moreover, it precedes the final definition of criteria for decision and selection-making. Introducing this change in the traditional (building) engineering processes and governance processes, where analysis and criteria definition are one of the first activities, is still not easy."

However, truly understanding and tackling these complex challenges remains difficult, because no single actor or organisation is, or can be, wholly responsible for them, while most aspects are interwoven and interdependent (e.g., Irwin, 2018; Van der Bijl Brouwer, 2022). These issues affect us all and in different spheres of life: as politician, voter, citizen, government official, business, technical professional, designer, or researcher (e.g., Smeenk, 2021). The collaborative and social-technological innovation and transformation processes that are needed to adequately meet these challenges are therefore dynamic, multi-stakeholder, and multi-sited (e.g., Kimbell, 2018; Vink et al., 2021).

The healthcare context, in which the third case is set, shows similar developments. Coming to actual implementation in collaborations between research and practice is still difficult (Gezondheidsraad, 2010). As an important way forward, there is a strong movement that advocates that clients participate as experts, under the motto 'nothing about us, without us' (Johansson, 2014). Furthermore, ongoing collaborations between researchers and practice partners are deemed important (Janssens, 2016).

What the above developments suggest is that an important process that we aim for in labs changing the status quo and working towards new ways of collaboration for sustainable common futures. Those involved need to be able to learn together. However, mutual learning is often assumed in participatory approaches, but rather taken for granted (Calvo, 2019). Pihkala & Karasti (2016) argue that more reflexivity is needed on the learning that takes place within the participatory design process.

In this chapter, we take a closer look at the learning that takes place within experimental environments, and we unravel the learning dynamics. We discuss three different case approaches to deal with this collaborative development challenge jointly. The three cases highlight the difficulties of finding a balance: to sufficiently address the multiple facets and voices, while also providing enough (externally expected) focus and momentum.

Theory: Boundary Crossing Learning Mechanisms

To discuss the dynamics in the learning processes between different collaborating partners, this chapter uses the four learning mechanisms which Akkerman & Bakker (2011) distinguish in boundary crossing processes as driven by the dialectic between different contexts. Table 1 summarises these mechanisms. Akkerman & Bakker distinguish two groups of mechanisms that focus on reflection and perspectives: *identification* of one's own identity and that of the other, and *reflection*, where those involved broaden their own perspective on the different ways of working on either side of the boundary. The two other mechanisms rather focus on activities: *coordination* of distributed work, in which those involved practically and efficiently coordinate their work, and *transformation* of previous ways of working and coming to new and hybrid ways of working.

Table 1. Four boundary crossing learning mechanisms (Akkerman & Bakker, 2011)

Identification of one's own identity and that of the other

- Othering, reconstructing the boundary
- Gaining insight in how practices differ
- Accepting differences

Reflection on the different ways of working on either side of the boundary

- Defining and exchanging perspectives
- Developing one's own perspective (perspective-making) and taking others' perspective (perspective-taking)

Coordination of distributed work

- Dealing with a boundary by each going their own way as much as possible
- Translating and communicating, aimed at efficiency in distributed work

Transformation of previous ways of working

- Joint work at the boundary
- Driven by mutual needs and a shared problem space
- Creation of new or hybrid forms

Moreover, Akkerman & Bakker (2011) note that boundary objects (Star, 1989) play an important role, especially in the mechanism *coordination*. They enable different groups to discuss and carry out their work. Between the people involved in the lab or experimental environment, a project proposal, memo, visionary visual or (paper) prototype can play such a boundary role. They are then boundary-negotiating artefacts, as indicated by Lee (2007). It is important to emphasise the fact that these negotiating artefacts can (and should!) be further developed into new versions of boundary objects. This dynamic is important to allow transformation of joint ways of working, moving beyond merely coordination, which is based on current, more separate approaches.

In the three described cases in this chapter, we will see that these dynamics play out differently, as well as the extent to which these boundary objects are seen as fixed, evolving, and immutable. The four mechanisms indicate how boundary crossing is more than just interacting with boundary objects. Between the people involved in the lab or experimental environment, a project proposal, memo, visual or prototype can play such a boundary role. All three cases show examples of – if sometimes temporary – consolidation of content of their joint work.

Moreover, it is about the process and group dynamics surrounding (further development of) those boundary objects. They then enable different groups to share, discuss, and carry out their work – even in a newly transformed way.

Dynamics in Three Cases

We present three cases which all differ on the topics of the intended transformation, but also in the roles we as authors were involved.

The first case concerns a participation project of a foundation that was set up with the goal to develop a new smart and sustainable district as a co-design process between future residents (starting with a new process among themselves), policymakers, building professionals, and other stakeholders. The case follows a co-design process of 2 years that started with an ambitious transition vision and no restricting formal conditions. Eventually, it turned out to be restricted by all sorts of unexpected governmental, ecological, economic, and social developments and interdependencies. The involved author had an explicit role of participation program manager (co-designer and facilitator) and facilitated the co-design process.

The second case concerns an energy transition program in a municipality in which the national, provincial, and municipal government collaborated with businesses to make their own real estate energy-neutral through a new area development approach. This case shows a clear tendency to focus content-wise on (joint) transformation and process-wise on (formal) coordination. The involved author had an explicit research role from which he tried to implement a design research methodology in the developing energy transition program.

The third case is set within two research programs by a Dutch funding party, aimed at enhancing long-term care while involving patients as well as practice partners. The experimental environment concerns the collaboration within the different consortia as they already start in the pre-project (definition) phase. Seven starting consortia used the same – new – organised way to jointly develop a research project proposal. The involved author had a role of facilitator of this new way of working and studied the process as design researcher.

We discuss all three cases along the four boundary crossing learning mechanisms. The results are summarised in Table 2.

Case 1: The smart and sustainable plot lab: an open promise

This case entails an ambitious urban planning program which aimed to create a hugely innovative smart and sustainable district, designed together with its various stakeholders, including future inhabitants. A foundation entity and accompanying program team was purposely set up, as a separate entity, apart from the city council and province, as well as universities and business in order not to be restricted by current traditional housing development ways of working, systems, procedures, etc., but to be more flexible, open, and actually work in an integrative way. A purposeful integral development program – explicitly not as separate projects – was set in place.

One of the authors was hired and assigned as a participation program manager to initiate the first experimental participative 'lab' environment for bottom-up 'co-building', and to facilitate the co-design process of the future residents; including communication between the future residents and the foundation. She, as a co-design expert, set up the participation process and aimed to empower future residents themselves to eventually take ownership and responsibility of the co-design process, as shown in Figure 1.

The assignment for this so-called *plot lab* included design of an open invitation in (social) media by an advertisement for an information evening (as usual when selling housing lots), with inspiring persona stories, explicitly inviting unknown ambitious *pioneering* future residents and house builders who are *willing to take risks* and be the first to live in the new, to be built smart and sustainable district. The pioneers were asked to co-design and co-develop a plot of land of 20.000 m^2 with approximately 40 households. There was no predefined urban plan and no planning. The restrictions only concerned seven innovative themes (1. sustainability, 2. mobility, 3. social and safe, 4. healthy, 5. participative, 6. data for residents, 7. energy neutral), and the expectation that the future residents take not only housing but also infrastructure along in their joint and individual plans.

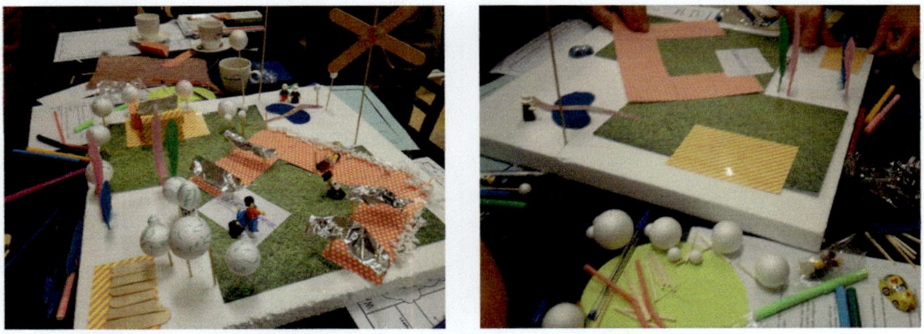

Figure 1. During the co-design process of their new neighbourhood, the involved residents developed mock-ups which functioned as boundary objects.

Identification

At the start of this lab, everything was open; the process and the housing plans were to be co-designed. At the first information evening, future residents were invited by the foundation director to collectively initiate their own experimental lab with support from the participation program manager as a co-designer. The factual information given to the residents was restricted to co-developing a plot of land of 20.000 m^2 with approximately 40 households, taking into account the ambitious seven themes and a fixed square metre price. Moreover, the future residents could decide themselves where they wanted their lots to be positioned and how the lots would fit together including a joint energy, water, mobility, and data infrastructure.

The first thing future residents warned for is that they did not want to be in an *'experiment'*. They emphasised they had a *real interest* to design, plan, build, and finance their aspired sustainable and smart futures, and that they were serious about realising their dreams with this once-in-a-lifetime opportunity. In the subsequent co-design sessions, all future residents spent a lot of time, effort, and risk (financial as well as social) to make this happen. They became a true community and felt they were trusted by the foundation program team and that they had the mandate to take the lead in planning and building their own neighbourhood.

Table 2. The topics to address within a collaboration when facilitating joint learning for the boundary-crossing learning mechanism (Akkerman & Bakker, 2011).

Topics to address	**Case 1:** **The smart and sustainable plot lab**
Identification of the own identity and that of the other **Deal with:** - Different views on knowledge and learning - Different views and unclarity on the design space - The balance of who is open to whom	+Focus on what residents want and can do as a value in itself +Between residents interests are shared transparently, which creates trust +Insight in three different areas of interest between residents - Other parties find residents' insights interesting, but do not share their own interests openly - (perceived) Power of the foundation makes residents suspicious
Reflection on the different ways of working on either side of the boundary **Deal with:** - Empathic formation - The time this takes	+Residents took time to exchange perspectives supported by appropriate creative methods, boundary objects and experts and made(shared) decisions - Lacking exchange and shared decision-making between foundation and residents
Coordination of distributed work **Deal with:** - Different views / expectations on boundary objects - Knowledge asymmetry - Capacity differences - Power differences - Temporality	+The neighbourhood as focal point and boundary object +Residents invest much time and effort, take risks, and learn quickly together +The commitment, time, and creativity by the designer as facilitator - Knowledge asymmetry - Speed residents faster than procedures of foundation
Transformation of previous ways of working **Deal with:** - (un)willingness or fear to step beyond usual ways of working - Temporary role of an external facilitator	+Designer dares to take responsibility for an open experimental learning process desired by the foundation +Idealistic residents take their mutual needs and role seriously - The foundation is accountable to the board. Designer is paid by the foundation. Residents are voluntary. So, accountability is not organised equally

	Case 2: **A developing transition program**	Case 3: **Circling in research**
	+Shared understanding of technical goals - No explicit interest in each other's learning preferences	+Focus on the multiple perspectives, on all project aspects - 'who should be at the table' is not explicitly addressed in the method - The identification process was limited to a relatively small core group
	+Many (informal) reflections occurring in sub-projects and sub-tasks - no consciously undertaken joint reflection	+Time was taken to really understand others' perspectives - Real empathy could still be better or more explicitly facilitated
	+New insights and innovations process-wise - Traditional division of already known work	+Working on a proposal as a boundary object, provides access to information for all +External facilitator to support the process - Hard to refrain from seeing the proposal as fixed, once accepted - Unaddressed capacity (and hence power) differences between project coordinator and the rest
	+Hands-on experience of working differently +Coupled to content development - Performed by hired market parties and only a small number of the involved governmental program partners	+Starting from a shared problem space - Not clear yet whether the learning dynamics, set in motion in this pre-project phase, will continue in actual project execution

Already by the third session, three future resident groups with similar ideas and wishes had emerged, namely, tiny houses, sustainable detached houses, and life-long collective buildings. Moreover, there were first ideas and plans on who would build close to whom, or together with each other, and how communal space would be developed and shared based on common needs and aspirations.

However, the interests and expectations of the foundation proved to be different from that of future residents. This became only apparent during the process. The participation of future residents resulted in concrete ideas and proposals that triggered the foundation to respond and also explicate its own (up until then implicit) expectations on aesthetics and quality. It set restrictive boundaries to the seven themes. Because the common purpose between future residents and the foundation suddenly seemed to be lacking, an inclusive co-design incorporating both groups and views never did (or had a chance to) happen.

A process of identification was visible among the future residents. They, in the spirit of true pioneers, were open to discovering what each other's views were, and to engaging in a joint learning process of who they are as a group, and what each individual wanted. Partly assisted by external architects, they were able to translate these new insights in joint proposals for the new housing development. They developed boundary objects and played with them. The same cannot be said for the foundation who needed to first experience what the boundary objects could be, based on which further decisions would be taken while not really engaging in joint learning or mutual identification of preferences.

Reflection

Everything was initially, or at least seemed to be, open. In the co-design sessions, future residents exchanged what they had in mind for their personal space and communal spaces. Most were attracted to the project by the seven idealistic themes, the overall communal aspect, the specific co-design process, and/or the prospect of building one's affordable own house. Their mindset was cooperative and necessarily reflective from the start. Facilitated by the participation program manager and her creative co-design methods (including boundary objects development and arranged expert exchange), they were challenged to exchange and find their differences and shared interests, knowledge, strengths, and aspirations regarding the seven ambitious and idealistic sustainable and smart living themes. This gave them a lot of time together to exchange perspectives, get to know each other, build trust, and eventually support and help each other with individual and collective plans to co-design and co-develop a common neighbourhood.

The foundation director and municipality architect were invited in explicit reflection meetings to exchange progress on both sides. The architect and director were critical on the quality and aesthetics of the ideas of the future residents, probably from their predefined (not known to others) set of criteria, but they did not always think along or give support in tips, tricks, or a network that could help the future residents. So, there was reflection among the future residents, but almost no constructive reflection between the residents, building professionals, the foundation, and the municipality.

Coordination

The organisation of this *plot lab* started off innovative, with a foundation entity that was purposely set up apart from existing structures such as politics, universities, and business so as to not be restricted by current (more closed) systems, procedures, etc. The 80 future residents were grateful for the participation program manager's support in facilitating the first sessions, making everyone feel welcome, and using creative co-design methods to make them thoroughly exchange their own and collective wishes and ideas. The foundation connected and organised governmental officials, policymakers, business experts, and researchers for support and information.

A positive result was that insights about the smart and sustainable living wishes of the future residents were obtained and shared within the foundation, which gave a new perspective on what people actually aspire and wish for in building new neighbourhoods. Yet, while developing their plans further at a rather fast pace, the three future resident groups ran into unclarity, difficulties, and process delays mainly because of other foundation and municipality processes lagging behind.

For example, the innovative zoning plan was not even started, the quality book not established, and the quality team not installed. These restrictive issues had a 'closing' effect on an 'open' experimental development process.

Eventually, coordination problems in temporalities and knowledge, information, and power asymmetries, led to communication and trust problems, irritations, mistakes, and finally deterioration and mistrust in the relations and dynamics between future residents and the foundation. The future residents depended on the way foundation professionals worked and vice versa. Residents felt that they were successful and fast moving and ready for the next step, but the foundation and its network partners were not. Many residents voiced their concerns about unclarities, first to the participation program manager and later directly to the foundation. At first, the participation program manager was able to further adapt the joint process in finding other relevant program elements to work on, share, and learn about. However, and since this (learning) coordination was not her sole responsibility, the participation program manager was unable to continue the co-design process and 'open' development of the project content.

Transformation

At the start of this lab experiment, the foundation program team, the participation program manager, and the future residents were enthusiastic and eager to work differently. This idea that the new district would be totally different, with true pioneers in the lead and with seven ambitious themes increased the enthusiasm that real transformation was possible.

However, by being fast and successful with this bottom-up co-design process with the future residents in the lead from the start, in addition to the foundation and municipality processes not up to speed, caused them to set off too quickly and then become overly enthusiastic. This caused the envisaged co-design approach to run into problems. Although some provisions were taken to be able to work outside the traditional systems, setting up a foundation and appointing responsible program managers, for example, the trailblazers were not able to keep that going, as they were not supported in their transformation process information demands. One could wonder if the setting up of a dedicated foundation was truly different compared to usual housing development approaches, since its director eventually had to comply with the rules and the usual actors in the current system. This resulted in the old roles and activities interfering more and more with the newly developed ones.

Discussion

In this example of a Living Lab, future residents were able to identify each other's housing preferences and wishes, reflect on them, and coordinate a joint plan. However, other import ant-actors like the foundation and the municipality were not able to critically reflect on their own role, coordinate the required efforts within their own organisations and network, and facilitate the transformations needed to make this lab plot a success.

The open process clashed with the emerging and gradual closed content requirements of the municipality which was outside the influence of the participants of the lab. For true transformation, the commitment and participation of these other parties' decision power was necessary. When this did not happen in time, the co-design and joint learning got stuck and in fact stopped, despite all (additional) efforts of the participation program manager.

One important factor was that for the future residents the innovation risks, especially financial and personal risks, were more direct and much higher than for the foundation and other actors. This meant that the interests of the future residents were much more oriented towards clarity and short-term 'closure' than the organisations who were risk-free interested in 'open' innovation. The future residents clearly had to deal with information asymmetry; much of the crucial information and expertise from the side of the other parties in the program organisation was not reaching them, which also prevented the other stakeholders from benefiting from the expertise of the future residents. Expertise and learning were not integrated.

Case 2: A developing transition program: a closed start

The second case concerns a joint (national, provincial, and municipal) government program. It aims to set an example of an energy transition approach by making governmental buildings in the centre of a major Dutch city climate neutral through a new area development approach which is reflected in a developing program consisting of sub-projects and tasks. The selected and essentially technically oriented consortium advises the governmental parties how and which innovative solutions to implement and integrate.

The main challenge of the program is how to *jointly* manage the developments. In other words, how can the developing energy transition program, in addition to being a joint transition tool for involved partners, also serve as an individual handle for their different goals while safeguarding commonality in energy transition interests. To this end, an existing design research methodology (Blessing & Chakrabarti, 2009) is used by the involved design researcher which follows an iterative designerly way of working. This design researcher is not in the lead of this process, also not as a dedicated facilitator, but consciously takes a distant and rather 'free' or 'open' role: he observes, asks questions, but also makes remarks and proposes interventions based on written information, sub-projects, and joint (co-design) meetups. His approach, which is approved by the program director and steering committee to help develop and demonstrate another way of working, is to go back and forth by discussing 'what is' (descriptive), 'what needs to be done' (prescriptive), and then to evaluate what the new situation is (descriptive) and move further. He hopes to eventually make himself redundant. The idea is that this new approach, which helps to reframe both processes and tasks, will be adopted during the program development and sub-project activities, and directly put in practice by the involved stakeholders. The approach therefore results not only in describing the course of program development, but also in translating it along the way into program characteristics.

Identification

In the initially tech solutions-focused consortium, the program team did not dedicate time at the start of, or during the program development to identify, note, or make explicit the way the involved individuals and organisations learn. This means that identification did not really occur on this aspect.

The assignment of the involved researcher is to regard learning from the perspective of an energy transition mission. As a design researcher, he is also personally convinced that we learn best while doing and by working together, even if it is implicit, and that we need to take questions and answer loops into account while innovating and implementing. In other words, he wants to transfer design research knowledge (attitude and skills) hands-on to practise in order to transform the collaborative work in a more integrative, flexible, and iterative open approach towards energy transition mission.

The technical development is based on a so-called 'Trias Territoria' working philosophy:

- Step 1: reduce the energy demand at building level, for example, through insulation, heat recovery, and energy-efficient lighting.
- Step 2: use and share local energy sources in the immediate vicinity, and determine what capacity they provide, and whether that capacity can be shared with other buildings in the area.
- Step 3: purchase sustainable energy from the region, such as heat, cold, electricity, or possibly hydrogen in the future.

An interesting development in the program was that, almost unnoticed, a group choice had been made to start first with the step two of Trias Territoria, or at least give priority to (joint) area measures, before choices for energy-saving measures at (individual) building level would be made.

This was a significant signal in the program, especially since one of the main explicitly stated goals was to learn in order to be able to repeat and pass on the developed solutions and new ways of working (technical, procedural, processes, rules, and regulations). What happened was that the explicit individual learning possibility presented by step 1 of the Trias Territoria approach was postponed and therefore not actively pursued by the program partners (governmental participants and owners). Instead, the development has been focused on step 2, innovation and implementation of joint solutions. The thinking was that this joint step 2 'hopefully' may prove sufficient enough to definitively avoid/skip the individual step 1.

However, what was hereby also unnoticedly skipped was the recognition of the own 'identity' regarding the task at hand (content-wise) and towards what is needed (process-wise) to meet this task. Instead of explicitly presenting and explaining to each other what the different perspectives and identities are, related to the possible actions regarding step 1, the shortcut of the available Trias Territoria step 2 (also literally focusing on the already available possible technical solutions) was taken. This way the true identity of 'the other', related to the task at hand, which was never really explicitly stated or consciously shared and discussed within the program team.

Reflection

The program team was willing and able to reflect, but rather on action than in action. The lack of deliberate identification meant that there was also no consciously undertaken joint reflection on the traditionally different ways of working by the program partners and involved stakeholders. Indirectly there were of course many (informal) reflections occurring in sub-projects and sub-tasks, often one-on-one, resulting in frequent improvement of understanding between the partners that were directly involved in a specific activity. However, leaving the rest of the program team often guessing what actually is taking place, and how it fits into implicitly already-taken decisions. The explanations offered during biweekly team meetings focus generally on technicalities of the proposed (sub)solutions, and what we learn from (trying to implement) them, but almost never on identity aspects or joint learning mechanisms.

Only afterwards are the efforts taken to (partly) describe what the lessons learned were and how certain results have been realised, in the so-called 'guidelines'. But even there it is explicitly stated, as one of the 'collaboration principles', that 'we are only looking at what connects us regarding the task at hand, and not what our differences are'.

Coordination

The learning mechanism of coordination, largely based on traditional division of already-known work between different disciplines in the sector, has led to new insights and innovations process-wise, specifically regarding procurement and contracting measures. This resulted in projects in which technical measures have been implemented.

Additionally, through reflection on joint ways of working regarding this type of familiar coordination and concerning regular program risks and subsequent measures, the whole program team follows a common learning curve directed to further process optimization.

One of the explicit program goals is the development of new ways to collaborate (that unfortunately often end in a process separate from content), and from a knowledge development perspective as well as a descriptive-prescriptive-descriptive perspective iterations were introduced and followed. This resulting learning effect was exactly wat was wanted. Coordination mechanisms did not however result or contribute to new energy-content innovations that drive the wanted transition.

Transformation

Transformation as a learning mechanism (coupled to the content development) is essentially performed by the hired consultants and market parties, and only by a small number of professionals from the involved governmental program partners. Since these governmental professionals act both as (1) clients to the hired professional consultants and construction companies, and (2) owners of the real estate that needs to be improved through energy transition measures, they assume a rather reactive attitude towards collaborative transformation and further (teach-the-teacher type of) learning within their own organisations.

To contribute to the energy transition, this program coalition needs to (experience and learn to) work differently individually and collectively. This transition program has as one of the aims to be an example of how to design, scale, and disseminate a transition process and knowledge to other practices and education. However, due to a focus on the more short-term sub-projects successes and deliverables, the learning aspect has not yet been fully realised and a new way of iteratively working, in a descriptive-prescriptive-descriptive shifting from traditional risk management to transformative new collaborative design opportunities, only partially occurred.

Discussion

From the perspective of four learning mechanisms (identification, reflection, coordination, and transformation), this energy transition program has developed a clear tendency along the way to focus only content-wise on (joint) transformation and process-wise on (formal) coordination.

Analysing the case using four learning mechanisms indicates that the identification (of one's own identity and that of the other) was insufficiently done up until now, and that this hindered the learning process. This is partly due to the (own felt) pressure to realise concrete (technical) measures, and partly due to the general avoidance of taking new energy transition measures on one's own individual real estate first. One of the resulting effects of this (partly implicit) approach is the creation of a largely closed type of experimentation, even though the program team and involved partners are aware of the fact that they are dealing with socio-technical (meaning one has to deal with multiple subjective and more complex interpretations) instead of only technical tasks (which could 'objectively' be functionally predefined).

The major consequence being that the buildings' end-users are not yet directly involved in the experimentation and (program) development, which is the same as the citizens who live in the concerned central city area. The other important consequence, which the program team is not yet fully aware of, is the continuous struggle to utilise the program as a joint transition tool that also serves as an individual handle for different goals of different partners, while safeguarding commonality in transition interests. In order to be able to accomplish this goal, one has to know what the (learning) identity of the involved partners and stakeholders is, to be able to fully utilise opening and closing dynamics for improved participation and collaborative development. It seems however as if in thistype of developing program, identification could be accomplished only after transformation and reflection.

A combination of a generally proactive joint involvement of the steering committee, together with a specific use of risk management as a development instead of as a monitoring and/or assessment tool, seems to offer some guidance and opportunities for prescriptive actions concerning further coordination, reflection, and transformation. But in order to open up more as a co-design and co-creation experimental environment, this program will have to find ways to foster identification learning mechanisms.

Case 3: Circling in research collaborations: to open up a closed start

The ‘Circling in research’ approach was developed within the context of practice-based research on healthcare innovations. This was in reaction to several problems in this context: research questions do not always fit actual practice issues, the inclusion of everyday people is not always satisfactory, and research results do not land in practice. Within two research programs in the field of healthcare innovations, a supportive method was developed and applied by seven partaking consortia to address these problems. The developed method is an organised way to iterate on decisions of aspects of a research project proposal, by circling around these decisions with a diverse team of stakeholders.

The circling method presupposes that involvement of all relevant stakeholders from the start strengthens the quality of the plan and its contribution (e.g., Johansson, 2014). Scientific knowledge, practice knowledge, and experiential knowledge are all treated as equally valuable in the method. Therefore, it is seen as important that the key stakeholders are present or represented at the table while circling. The attempt was to make the seven projects open environments as to who joins the table (Jones, 2018).

The circling process distinguishes several key project facets: practice issue, knowledge gap, research question, project approach, project conditions, goals, and products. The goals explicitly include not only knowledge goals. The circling approach is based on the four goals which Greven & Andriessen (2019) distinguish in practice-based research: knowledge development, product development, system development, and personal development. As a consequence, not only a research approach needs to be developed, but also approaches for the three other goals.

In the two research programs, the application of the method was facilitated by the project team which developed the method. One of the authors of this chapter was part of this team. The seven consortia that tested CIRC can be seen as seven Living Labs that have the intention to develop and execute a research proposal. Each consortium consists of several stakeholders including clients, caretakers, doctors, and managers.

Identification

One of the underlying design principles of circling is to foster a multi-voiced process, in which different perspectives do not have to merge and in which different goals and expectations can coexist. Special attention is paid to the process of meeting each other not only as a professional but also as a person. This stimulates bonding and the process of identification. One of the key findings was that the CIRC approach helped almost all of about 50 people who participated in the seven consortia to learn about the perspectives of others. For instance, by the participation of clients, nurses and doctors learned to see problems from their perspective.

However, the mutual *identification* process was limited to a restricted group within each forming consortium. Although CIRC promotes an open process for the development of research proposals, the seven groups which developed research proposals all included a rather limited group for practical reasons. Because the workshops had to be online, there was a maximum number of participants with which CIRC could facilitate a good dialogue. The consortia found several ways to address this limitation. Some held additional focus groups with stakeholders, and others formed a steering committee to give additional feedback. In some cases additional stakeholder groups were identified and invited to join in the research proposal.

Reflection

The challenge that is addressed in the circling method is how to make sure that these different voices are not only present at the table but *really heard*. The circling process helped to go beyond identifying different perspectives. It helped to understand one another and to be able to take another's perspective, for instance the different perspectives on what is 'knowledge' or 'good research'. Several elements of the approach are aimed to explicate underlying values. Some come from arts-based research, such as the contemplative dialogue (*www.musework.nl)* which fosters a process in which participants listen to each other very carefully. Still, the evaluations pointed out that real empathy between partners, and a safe space, could be even further stimulated.

Participants recognized this process of taking each other's perspective as being very helpful in developing the research proposal. However, what they did find difficult is that such reflection requires slowing down the process at some points. Slowing down was perceived as a challenge, as the consortia felt the time pressure to submit a proposal before a deadline. It helped that the participants in these two research programs were given dedicated time for this process. When they worked with the approach, they actually appreciated the slowing down and indicated that they indeed broadened their own perspectives.

Coordination

Central in the circling process is the development of a research proposal. This is developed as a prototype of the collaboration and can be seen as a boundary object in a *coordination* process. The prototype is an important way to provide clarity to all involved about the project, among which to the research funder, as it opens up information to the different stakeholders.

Not all seven consortia used the draft research proposal as a boundary object within their consortium. The ones that did were more successful in gaining support for the proposal in all stages of the process. The successful projects gave an update on the progress at the start of each meeting and created a document describing all information gathered and decisions taken.

The approach alternatively addresses the key project facets as described before. These are iterated upon instead of dealt with in a linear process. A certain logical route is presupposed in this, starting from practice issues and moving to goals and eventually a research approach. A process may start with a different step, for instance with a particular envisioned method. Participants will then need to loop back to check what the practice issue is, whether there is a knowledge gap, etc. Participants indicated that the process of 'circling' was helpful in postponing judgement and looking at the topic from different angles. However, it also made it more difficult to jointly decide on the focus of the proposal. Discussions were sometimes experienced as too broad and abstract.

Although the method promotes equality amongst partners, in practice there is often one party as the main supplicant. This introduces a tension to this equality and power differences. The other partners in a consortium (e.g., caretakers or clients) do not have the same time to extensively take part in the preparation. The circling approach prescribes suggestions for process steps in which a balance is sought between actions by this main coordinator and joint actions by all partners. Tips are provided for the process facilitator, to be process-sensitive.

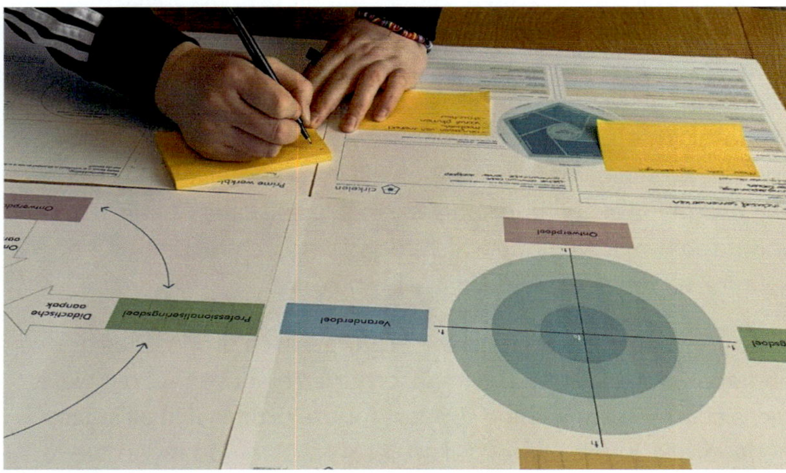

Figure 2. The canvas which is used in the 'circling in research' approach to capture the insights during the process and develop the research proposal.

What proved challenging is that not everybody can be at the table, especially in the large and layered organisations which the healthcare organisations in these projects are. And even when people from different layers in an organisation are all present, power differences play a role. Although the method is intended to be helpful to this end, e.g., by using methods to build mutual respect and by paying attention to personal relations, the matter of power differences remains underwater and is not explicitly addressed. The evaluations showed that 'who is at the table' should be one of the key aspects to address and discuss as part of the project proposal under construction, including the topic of the related power differences. This was initially not part of the key canvas, but was inspired by the Co-Design Canvas by Smeenk et al., 2023, and thus it is now being integrated as part of it to make sure it is a returning topic, which can change during the process.

Transformation

One of the goals of introducing the circling approach was to stimulate a different way of looking at scientific research in which scientific knowledge, practice knowledge, and experiential knowledge are treated as equally valuable, both within the participating consortia as within the funding agency. We are now in the process of researching whether this new way of looking at research, as it was stimulated during the facilitated process, remains present during the execution of the research proposals. We did see a positive change with members of the funding agency who are now advocating the inclusive way of developing research proposals, as promoted within the circling approach, further within their organisation.

Discussion

This case shows how joint learning processes in practice-research collaborations can be supported by a conscious effort to open up the decision-making and learning process. It shows how it helps to explicitly dedicate time and effort to this process.

In the current way the circling process was executed, the focus was rather on *identification* (multi-voicedness), *reflection* (how can we really learn from others), as well as *coordination* (the project proposal as central object). We do not yet know the effect on *transformation*. The learning mechanism of transformation draws from the power of working together from a shared drive and by coming to new ways of working: daring to change or put aside your current ways of working also as researchers.

We already addressed that research projects have an inherent power difference built in. The coordinating party (mostly the researchers) have more time and resources to give direction to the project, whereas practice parties often have limited ways to steer the project. While the circling process is facilitated by an external facilitator, as in our case, these facilitators also have power to steer the process towards a certain way of working and learning together. The challenge for the circling method is to help a consortium to come up with their own, unique, and really shared way of working, fitting for the context and stakeholders involved. How can the canvas or other elements of the method support this?

Conclusion

Collaborations in experimental learning, innovation, and transition environments such as Living Labs can be complicated to carry out in practice. In this chapter, we reflected on the opening and closing dynamics within three cases of such collaborations by using the lens of the four learning mechanisms in boundary crossing (Akkerman & Bakker, 2011).

Our three cases all show how we, as authors, took a facilitating role as designer, design researcher, and facilitator in collaborative experimental environments or labs. They illustrate how we are all driven, in these different contexts, by the need to (re)shape these environments in such a way that the involved actors can adopt new ways of working where joint learning can thrive. In the built environment as well as in the healthcare context, we recognize tendencies to adopt rather classic views of what 'knowledge' is. In this view, knowledge is viewed more as a product, whereas we also recognize the importance of viewing knowledge also as a process (Andriessen, 2008). In that light, the opening and closing dynamics we identify in experimental environments are all viewed by us in the light of *joint learning*.

All three of us, in more or less explicit facilitator roles, aimed to shape collaborations towards a new way of working which facilitates joint learning. This stems from our shared belief that the experimental nature of a lab or experimental environment should be about the whole process, about continuous learning and development of multiple stakeholders. We strive for change in the process to – in the end – attain different types of results and impact.

Looking back on our cases, we realise that we see it as important that (1) we, as such facilitators, reflect and do introspection on these – sometimes unspoken – ambitions of our own, and that (2) we share and further develop the ways of working which help to attain these ambitions.

In light of the first point, we propose that these ambitions can be too high. All three of us, therefore, are looking for ways to make ourselves redundant in time. This is probably for the best. As there is a risk of moving from missionary to table banger, there could be a point where someone else may better step in. This means that during the process, design researchers need to balance their role. They are often involved in a – more or less explicit – role as facilitator. In this role, they need to beware to be overly present and indispensable in the process. Whereas they also need to beware of getting too much on the sideline, becoming 'already redundant'.

For the second point, by viewing the three cases in light of the four learning mechanisms, we conclude that a lack of attention to any of these is able to hinder the learning process. For instance, the developing energy transition program case illustrates how a learning transformation really requires a process in which the different partners understand one another's needs and interests (identification). Therefore, we conclude that all four learning mechanisms should be considered in any lab and given due attention in the specific context.

Within each mechanism, we gained an overview of attention-worthy aspects when it comes to opening up the learning process. The three cases show that there is more to opening up a lab than just 'who is involved'. For instance, the plot lab case shows how joint learning can get hindered and a lab can get stuck, even when there is an enthusiastic group who identify each other's preferences and wishes, reflect on them, and coordinate a joint plan. We recognize that experimental environments take all kinds of decisions along the way which affect the learning dynamics and the learning transformation. The three cases together provide a preliminary overview of topics which collaborating parties should address, such as different views on goals, design space, and decision power (the first column in Table 2). We can imagine that experiences from other cases can further add to this list. Ideally, the relevant actors should be involved in decisions on these topics which affect the learning dynamics. This means that explicit, joint, and periodic reflection on all these topics is needed. The circling case provides an example in which this is attempted from the early start-up of a collaboration.

We propose that the insights from the three separate cases, as well as the overarching insights, can be helpful for other designers, researchers and stakeholders who aim to start an experimental environment or a lab and or take a facilitating role.

Bibliography

- Andriessen, D. G. (2008). Stuff or love? How metaphors direct our efforts to manage knowledge in organizations. *Knowledge Management Research & Practice, 6*(1), 5-12.
- Akkerman, S. F., & Bakker, A. (2011). Boundary Crossing and Boundary Objects. *Review of Educational Research, 81*(2), 132-169.
- Blessing, L. T. M., & Chakrabarti, A. (2009). DRM, a Design Research Methodology. Springer London.
- Dorst, K., & Cross, N. (2001). Creativity in the design process: co-evolution of problem-solution. *Design Studies, 22*(5), 425-437.
- Calvo, M. (2019). Co-design and informal-mutual learning: a context-based study demystified using cultural-historical activity theory. Doctoral Dissertation, The Glasgow School of Art.
- Chen, D.-S., Cheng, L.-L., Hummels, C., & Koskinen, I. (2016). Social design: An introduction. *International Journal of Design, 10*(1), 1-5.
- Gezondheidsraad. (2010). *Gezondheidsonderzoek aan hogescholen*. Den Haag: Gezondheidsraad.
- Irwin, T. (2018). The emerging transition design approach. *Cuadernos del Centro de Estudios en Diseño y Comunicación. Ensayos(73)*, 147-179.
- Janssens, M. (2016). Nut en noodzaak van netwerken voor de verbinding tussen onderzoek en praktijk. Den Haag.
- Johansson, V. (2014). From Subjects to Experts-On the Current Transition of Patient Participation in Research. *American Journal of Bioethics, 14*(6), 29–31.
- Jones, P. (2018). Contexts of Co-creation: Designing with System Stakeholders. In Jones, P., & Kijima, K. (Eds.), *Systemic Design*. Translational Systems Sciences, vol 8. Springer, Tokyo. https://doi.org/10.1007/978-4-431-55639-8_1
- Kimbell, L. (2018). How can we...? Connecting inventive social research with social and government innovation. In Marres, N., Guggenheim, M., Wilkie, A. (Eds.), *Inventing the Social* (Ch. 13), Mattering Press. https://doi.org/10.28938/9780995527768
- Lee, C. P. (2007). Boundary Negotiating Artifacts: Unbinding the Routine of Boundary Objects and Embracing Chaos in Collaborative Work. *Comput Supported Coop Work, 16*, 307–339. https://doi.org/10.1007/s10606-007-9044-5

- Lee, J. J., Jaatinen, M., Salmi, A., Mattelmäki, T., Smeds, R., & Holopainen, M. (2018). Design choices framework for co-creation projects. *International Journal of Design, 12*(2).
- Pihkala, S., & Karasti, H. (2016). Reflexive engagement: Enacting reflexivity in design and for 'participation in plural'. In *Proceedings of the 14th Participatory Design Conference, Vol. 1*, 21–30. Aarhus, Denmark.
- Sanders, EBN., & Stappers, PJ. (2008). Co-creation and the new landscapes of design. *CoDesign: International Journal of CoCreation in Design and the Arts, 4*(1), 5-18.
- Savanović, P. (2021). Integral development of the built environment. In Joore, P., Stompff, G., & van der Eijnde, J. (Eds.), *Applied Design Research – A Mosaic of 22 Examples, Experiences and Interpretations Focussing on Bridging the Gap between Practice and Academics*. Taylor & Francis Group.
- Smeenk, W. (2021). The world tilted. *Inholland, Diemen*.
- Smeenk, W. (2023a). The Co-Design Canvas, a proven tool for societal impact. *BIS publishers, Amsterdam*.
- Smeenk, W. (2023b). The empathic Co-Design Canvas: A tool for supporting multi-stakeholder co-design processes. *International Journal of Design, 17*(y), 81-98. doi: 10.57698/v17i2.05
- Star, S. L. (1989). The Structure of Ill-Structured Solutions: Boundary Objects and Heterogeneous Distributed Problem Solving. In Gasser, L. & Huhns, M. (Eds.), *Readings in Distributed Artificial Intelligence*. San Mateo, CA: Morgan Kaufmann, 37-54.
- Van der Bijl-Brouwer, M. (2022). Service designing for human relationships to positively enable social systemic change. *International Journal of Design, 16*(1), 23.
- Vink, J., Wetter-Edman, K., & Koskela-Huotari, K. (2021). Designerly approaches for catalyzing change in social systems: A social structures approach. *She Ji: The Journal of Design, Economics, and Innovation, 7*(2), 242-261.

Koen van Turnhout, Daan Andriessen

7. Experimenting with Novel Knowledge: a Plea for Communities of Practice

DOI: 10.1201/9781003491484

Introduction

Living Labs, Fieldlabs, and Urban Innovation Labs are environments in which participants develop new viewpoints on and alternative approaches for a local problem. The aim is that participants appropriate their diverse sets of knowledge for resolving the problem at hand. In this contribution, however, we'd like to focus on experimental environments where the production of knowledge that transcends particular problems is the primary goal. We will argue that Communities of practice (CoPs) can play a role in generating such transferable knowledge. Communities of practice are experimental learning and innovation environments in which participants collectively reflect on their experience in working on their own local problem, with the double aim of individual learning and the common development of new knowledge.

We argue that working in communities of practice has the potential of joint knowledge development but that it also offers several challenges. We illustrate this using two case studies. The first case study is the Design Science Research Group which was founded in 2006 as a community of practice for researchers interested in the methodology of Design Science Research (DSR), a methodology for the development of design-relevant knowledge in social sciences (Van Aken et al., 2016; Van Turnhout et al., 2023). The aim of the community is to create interest among researchers for DSR, stimulate individual learning about the methodology and collectively develop this methodology further.

The second case study is the Workplace for Musal Research, a nine-day gathering of researchers, artists, and practitioners working together on the idea of 'the musal perspective' (Van Rosmalen, 2016). This is a perspective in research and professional work inspired by the Greek Muses. The word 'musal' is chosen as an alternative word for 'art', a word that has a specific connotation but that does refer to the use of artistic way of working and thinking and artistry in professional life and scientific research. Participants in this community work on applying musal (research) methods in their own organisational contexts.

Both cases were chosen because each community aims for the development of new knowledge, in addition to the goal of personal learning. However, there are important differences between both cases, including different views on the nature of knowledge itself, that can shed light on some of the challenges when organising a community of practice for knowledge development.

This chapter is structured as follows. First, we briefly introduce the concept of communities of practice using literature from Wenger et al., (2002) and Ropes (2010). Then we describe the Design Science Research Group (DSRG) and the Workplace for Musal Research (WMR). We conclude with a cross-case analysis of both cases in which we try to answer the question: in what way can communities of practice facilitate the collective construction of new knowledge for practice?

Communities of Practice

Most of the articles in this volume are situated in the notion of Living Labs, Fieldlabs, and Urban Innovation Labs. These are environments in which participants develop new viewpoints on alternative approaches for local problems (Overdiek & Geerts, 2023). Often multidisciplinary collaboration is seen as a means to innovate practical contexts. The idea is that participants appropriate and synthesise knowledge from diverse (professional) perspectives to remedy a situation that is seen as problematic. However, in this chapter we are more concerned how such environments can be better equipped to develop knowledge that can be reused in multiple contexts: transferable knowledge. We define transferable knowledge as knowledge that can be shared across individuals and can be appropriated for many different situations. The notion of communities of practice (CoPs) is suited for this case, because these are environments in which the commonality of the participants is not organised around the local problem but around the topic of interest. A CoP is 'a learning partnership among people who find it useful to learn from and with each other about a particular domain' (Wenger et al., 2011). Participants in a CoP use each other's experiences in individual practices as a learning resource and join forces in sense-making and addressing challenges they face individually. CoPs are not organised around a specific societal problem or challenge, but around a theoretical concern or a body of knowledge.

A CoP consists of three elements (Wenger et al., 2002): a domain of knowledge that inspires members to participate, a community that creates a social fabric for learning, and the individual practices of the members. The purpose is to learn together about the knowledge domain during social events like open dialogue, presentations, and exercises and apply this knowledge in the local practices. An additional way to generate value can be to jointly develop new knowledge; knowledge that is new to the world. It is this generative ability of CoPs that is the focus of our article.

The value creation of CoPs is multi-layered. Wenger et al. (2011) distinguish five cycles of value creation:

1. Immediate value creation in the activities and interactions. This is the case when members help each other during gatherings by providing tips or sharing experiences.
2. Potential value creation through the creation of knowledge that can be used at a later stage.
3. Applied value creation through changes in practice that are the result of the application of this knowledge.
4. Realised value through the improvement of performance that results from these changes.
5. Reframing value when the learning causes a redefinition of what is successful in a local practice.

The focus of this article is on the second cycle in the list: potential value creation that can be used at a later stage. What is interesting to note though is that Wenger et al. (2011) see the knowledge that is being created primarily as individual knowledge and not as generic knowledge that transcends the specific context in which it has been developed. They state that CoP activities and interaction can create 'knowledge capital' that can take different forms:

- Human capital: personal assets like a useful skill, a piece of information, or a new perspective
- Social capital: relationships and connections
- Tangible capital: resources like information, documents, and tools
- Reputational capital: collective intangible assets like the reputation of the network
- Learning capital: a transformed ability to learn

What is not mentioned in this list is the creation of knowledge as conceptual artefacts (Bereiter, 2002). Conceptual artefacts are intellectual constructs that can be used to represent, understand, or communicate abstract concepts, ideas, or knowledge that are shared among multiple people, that can be transferred through publications and tools, that transcend multiple application contexts, and that have a particular quality because they are validated. CoPs can be vehicles to produce conceptual artefacts. Andriessen (2003) quoted in Dekkers et al. (2005) described this function of CoP as the third stage in the development of a CoP, the three stages being (1) individual knowledge sharing, (2) joint knowledge codification, and (3) joint development of new knowledge.

Figure 1. One of the working methods used at the Workplace for Musal Research is the 'collective poem'.

However, reaching this third stage is not easy. In two case studies we will study how this was achieved and what dilemmas were encountered.

Design Science Research Group

The initiative

The Design Science Research Group (DSRG) was initiated by Joan van Aken and Daan Andriessen in 2006 in order to stimulate and develop the design science research methodology within the social domain, at the time at a nascent stage. Design science research engages in the design and evaluation of real-world problems in order to generate scientifically validated, design relevant knowledge that is usable in multiple contexts (Van Turnhout et al., 2023, p.20). Nowadays, design science research is part of a rich family of design-based approaches to scientific research, described and compared in Andriessen & Van Turnhout (2023).

The idea of the DSRG was to assemble a collective of practitioners of design science research (in their own practical domain such as management science or educational sciences) that were willing to reflect on their experiences in order to address issues with and jointly improve their methodology. The group held regular meetings in which researchers shared dilemmas, cases, and issues related to the methodology. Initially the frequency of meetings was quite high (monthly), but over the years it dropped to a quarterly frequency. Nowadays, meetings are primarily focused around overarching issues for which the individual experience of the group members provide valuable case materials.

The results

Members visit meetings regularly to get inspiration and valuable contacts for their own research; one might assume that this also improves their research practices, although such outcomes are hard to quantify. The meetings of the CoP have resulted in new knowledge that has been disseminated in the form of a handbook that is considered an authoritative resource in this field (Van Aken & Andriessen, 2011). Recently a new version of the handbook has been published incorporating many new insights that emerged in the CoP after the publication of the first handbook (Van Turnhout et al., 2023), demonstrating the sustained power of the community to generate new knowledge.

Enabling factors and dilemmas

The DSRG is a strong community bound by a shared interest in the methodology of design science research. Its success as a long-lasting community of practice is partly due to a steady influx of novel members who want to learn more about the methodology of their choice, and a solid core of long-term members seeking a platform to discuss methodological issues. This mix enables novice members to grow and gradually take more agency and a more vocal role in the community (see Fischer, G., 2011); experienced members need to make sure their discussions remain accessible to novice members, which ensures that novel insights remain accessible to novices, but which may limit the speed of developing new insights.

The core of the com munity is formed by a group of participants who jointly fetch a year's program and organise workshops. This core group typically consists of more experienced members, which ensures that long-term development of ideas is a consideration in setting the agenda for the group. A workshop typically is prepared by organisers by reading up on a topic, or consulting experts, addressing a core question with the participants (based on their different practices), and making some sort of a report after the meeting. So, workshop organisers learn in three stages, whereas other members mainly rely on the workshop of a learning opportunity.

The combination of a shared interest within a diversity of practices is a strong element of the approach. Each meeting, participants share experiences from a wide range of contexts and reflect on them from a certain overarching theme. This enables members of the community to grow and meeting organisers to cross-validate their insights with a number of practices (Khan & VanWynsberghe, 2008). This way meeting organisers have an opportunity to validate the applicability of their insights across a wide range of practices.

It must be noted that these validations are not always as stringent as one would like them to be. Partly this is because of intrinsic limitations in the validation of methodology, partly because of limitations in how a practice is represented by the participants of the workshop. Only so much can be brought to the table in a 2-hour session, and as such the actual practice which is reflected upon may be lost in translation. Moreover, the organisation is more suited to support the inductive process of looking at one's own practice in a novel way, than the deductive process of testing novel insights in practice.

Workplace for Musal Research

The initiative

The Workplace for Musal Research (WMR) was initiated in 2017 by Bart van Rosmalen and Peter Rombouts from HKU University of the Arts Utrecht, in cooperation with the Universities of Applied Sciences Utrecht and Rotterdam. The goal is to create a space for participants to develop different ways of thinking and acting based on principles derived from the arts. To avoid the strong associations many people have with 'arts', the initiators adopt the work of Van Rosmalen (2016) who goes back to the founding mothers of both the arts and the sciences, the nine Greek Muses. Every year a group of about 20 people joins nine times, in one-day or two-day sessions, to work on issues from their own practice. Participants included practitioners, artists, and researchers from a variety of organisations. Not only do they develop ideas and solutions for their own problems, but together they work on new ways of collaborating and working. Important in the ways of working within the workplace is the constant combination of ethos, logos, and pathos (Rombouts, 2023): working on the ethical dimension of the problem, the factual dimension and the ways to reach the hearts of the people involved. Like in a real workplace, the group not only talks about the issues but also creates artefacts, ways of working and works of art that can help with the issue at hand. Together, the groups try to research these ways of working and the principles behind them.

Figure 2. Association cards used at the Workplace for Musal Research.

The results

Each member is encouraged to bring a problem or difficult task from their own practice, work on it together with other participants during the sessions, and experiment with the ideas in their own practice. During the sessions, there is a lively interaction among participants. Because the ways of working developed in the workplace are not only aimed at the head but also at the heart, members often are deeply touched by their experiences and inspired to try new things in their own practice. They go home with new ideas, try them out, and bring their findings to the next session. We did not systematically research the impact, but our impression is that in many cases the experiences in the workplace have changed lives and practices.

Members are encouraged to document their findings in blog posts that are gathered in a website (*www.musework.nl*). Furthermore, for 2 years in a row, each member was asked to write a piece for a joint bundle called 'Blad' (which means both 'magazine' and 'leaf') about their project and their experiences and to organise a session during the closing festival.

Enabling factors and dilemmas

The Workplace for Musal Research (WMR) had a strong impact on its participants. Members report that they learn a lot about different ways of working and ways to integrate an artistic approach in their daily work. Its success is partly due to the well-prepared and designed sessions and a collection of strong work forms that address head, hands, as well as hearts. The facilitators are able to set the right scene by carefully choosing the environment in which the sessions take place as well as striking the right tone in the kick-off of each session. The sense of community is further stimulated by having a meeting with each participant individually before the workplace takes off to discuss personal learning goals and opportunities to bring the learning into practice in their own environment.

Joint knowledge production about the common denominator of the community: the musal way of working, is, however, more difficult for this community. The contributions of the participants in the joint publications are mainly individual expressions and reflections and not the result of a participative process of sense-making. The facilitators of the workplace are the primary creators of such new knowledge, although there is a tendency among the facilitators to start each year anew instead of accumulating past experiences (Andriessen, 2022). Although there was ample opportunity to jointly develop knowledge about the effects and application of prototypes of work forms, not much systematic reflection took place on the experiences of participants using those work forms in their own practice. The second author of this chapter has been part of the facilitating team of the workplace. Being an expert on methodology, he made several attempts to structure the process of collective knowledge development. For example, twice he has introduced a model that describes the steps towards knowledge development in the community, but both times the model was only used for a short period of time (Andriessen, 2022).

What may be reflected here is a difference in epistemic culture (Knorr-Cetina, 2007) between the artists in the community and the researchers. The researchers, who were primarily educated in the social sciences, viewed knowledge as justified generic beliefs, while the artists viewed knowledge as temporal inspiration unique for a local situation. When the researchers tried to structure collective experiences into generic findings (for example, by creating a model), there was a fear among the artists for reductionism and simplification. What may play a role here is that many artists consider it their role in society not to summarise but to disrupt reality. Structuring reality is dangerous, as it consolidates existing ideas while it is the task of the arts to challenge existing ideas. This 'rebellious' behaviour impeded (in the eyes of the second author) the joint construction of new knowledge.

Lessons Learned across Cases

In this reflective section, we compare both cases in order to draw out lessons about how a community of practice can be utilised to create new knowledge that can be transferred to multiple contexts. In the next section, we briefly speculate about how these lessons can be implemented in Living Labs, which do not have transferable knowledge as a primary goal.

Lesson 1: CoPs are strong instruments for individual learning

Both in the DSRG and the Workplace for Musal Research (WMR), most participants have a strong individual motivation to participate. DSRG participants are practising researchers who think their engagement with like-minded people can strengthen their research. In the WMR, participants want to learn about new forms of work to transform their practices. In both cases a sense of belonging to a community of like-minded people acts as a strong motivator. DSRG does not monitor individual learning, but many participants of WMR report the community gives transformative learning experiences for them and their teams. It is also something that the WMR organisers strongly encourage and foster through an intake and regular checks on individual learning.

Lesson 2: The core challenge is to bring the CoP beyond individual learning experiences and collectively work on generic knowledge development

CoPs can be used for generic knowledge development by combining local experiences brought in by participants. They all have their own practice and use the CoP to reflect on it. Harvesting these reflections requires systematic analysis of communalities in local situations and problems and co-designing possible solutions. The DSRG and WMR feature different approaches for this systematic analysis. The DSRG approach could be described as top-down. A knowledge agenda from a core group sets the agenda for knowledge development in the community. Each meeting is centred around reflection topics, individual experiences from the members are only considered with respect to how they relate to the theoretical notion that governs the meeting. The WMR approach is much more bottom-up; individual experiences are central to the meetings and the theoretical notions that are discussed are invoked because of their relevance to the case at hand. In the DSRG it is easier to develop generic knowledge, as this is the starting point, but the drawback is that the relevance to the practices of the participants may be weak. Whereas in the WMR the individual participants benefit, but there are challenges in combining the different experiences into more generic concepts.

Lesson 3: Different approaches of organising a CoP afford different forms of validation

The DSRG and WMR also have marked differences in how they facilitate validation. The top-down approach of the DSRG gives ample opportunity for cross-case validation. In the sessions participants from different contexts apply the ideas. As such, workshop organisers can get a good sense of how applicable their ideas are in different contexts. However, in the DSRG 'everything' happens in the session; so there is no callback from members applying concepts in practice and giving feedback about the utility of the concepts there. This practice validation is at the core of WMR. Participants get an intake, develop an intervention plan for their own practice, and make personal learning goals. However, there is no systematic process for collecting the individual experiences. Ideally new knowledge development would incorporate both practice and cross-case validation; so a hybrid approach may be beneficial to both communities.

Figure 3. Group work at the Workplace for Musal Research.

Lesson 4: The creation of joint knowledge products may facilitate knowledge production

The joint development of new knowledge can be boosted by working on concrete outputs, like a book. In the DSRG the two handbooks have been a major vessel to share knowledge and to foster the influx of the community. The WMR has been creating collections of individual contributions to share inspiration and to create a sense of community among participants. There is also a lot of attention for making concrete deliverables in the WMR. However, the DSRG handbooks have the ambition to communicate a coherent vision, whereas the WMR publications are more collections of individual expressions. Integrating experiences and creating common knowledge takes substantial effort, such as discussing key concepts in the group and having an editorial team that guards the vision and coaches the contributors in their writing process.

Lesson 5: CoPs need to foster different levels and types of participation

Both communities have a space for different participation levels and types (this is considered good practice by Fischer et al., 2011). In the DSRG a coregroup organises the activities and sets the agenda, a group of frequent visitors co-shapes the ideas, and a corona of interested people occasionally step in. During their engagement with the DSRG, members step up or down in participation level. The WMR has a fixed core group of organisers, and a more coherent group of members that work together for a year. A lot of care is taken in the WMR to cast the members, which ensures diversity of and synergy between participants, and to come to a mutual 'contract' of participation. Organisers of CoPs need to think through what individual motivations may be to join the community, how these can live up to the overarching goals of the community, and how participation levels can be stimulated.

Lesson 6: Different epistemic cultures require different ways of organising CoPs

The DSRG and WMR have very different epistemic cultures (Knorr-Cetina, 2007). The top-down, reflection-centred approach of the DSRG may be tailored towards social scientists who are most comfortable in taking distance from a situation and codifying it in explicit language. A substantial part of the WMR participants are more comfortable in seeing knowledge as grounded in and having unique relevance to personal experiences. Both communities differ in their ideas about the character, status, and preferred embodiment of joint knowledge. Such implicit ideas on what knowledge is, how it can be created, and (re)used ask for different ways of working and raise different challenges in fostering knowledge creation.

Conclusion: Integrating CoPs in Experimental Learning Environments

Most Living Labs are rich environments where a diverse group of stakeholders is organised around a certain problem (or set of problems). Learning within the lab is organised accordingly. Participants from different disciplines learn from each other to be able to solve the problem in a better way. More often than not, reusing that knowledge beyond the lab is not the main concern of the participants or the organisers. In this chapter we argued that this might be a missed opportunity. Often the novel insights generated in a labsetting can be translated into more durable knowledge and the organisation of the lab could facilitate this.

Communities of practices can be a starting point to think about such explicit knowledge creation. One could form guilds within the lab or across multiple labs of people who have similar, more overarching learning questions and different contexts to relate the lessons from the lab to. Within the Netherlands, the funding agency NRPO SIA has initiated a funding scheme for research infrastructures called 'Sprong'. This infrastructure can be used to facilitate such learning 'across' Living Labs. For example, the Sprong ESC (Expertise Network Systemic Co-design) is an infrastructure around systemic co-design research, and it is used to organise exchanges between researchers in different labs (Smeenk, 2021). At current, exchanges between 'specialists' across labs already take place, but these have a 'pop-up' character.

Our analysis suggests a more sustained effort with a stable group of people may be of benefit to this research infrastructure. In order to make such a structure a success, one needs to consider individual learning needs and how these can be matched, collective activities that foster joint reflection and knowledge production, the way in which these insights can be (cross)validated, what knowledge products are feasible outcomes, how different levels of participation can be facilitated and intersect, and what epistemic cultures are needed to make the CoP a success.

Acknowledgements

The authors would like to express their gratitude to Wina Smeenk and Stefanie Bumbacher for their valuable contributions to this chapter.

Bibliography

- Van Aken, J., Chandrasekaran, A., & Halman, J. (2016). Conducting and publishing design science research: Inaugural essay of the design science department of the Journal of Operations Management. *Journal of Operations Management, 47–48*, 1–8.
- Van Aken, J., & Andriessen, D. (2011). *Handboek ontwerpgericht wetenschappelijk onderzoek: wetenschap met effect*. Boom Uitgeverij.
- Andriessen, D. (2022). *De taal voorbij*. Musework. Retrieved from https://www.musework.nl/nl/page/12097/de-taal-voorbij.
- Andriessen, D., & van Turnhout, K. (2023). Stromingen in ontwerpgericht onderzoek. In van Turnhout, K., Andriessen, D., & Cremers, P.H.M. (Eds.), *Handboek ontwerpgericht wetenschappelijk onderzoek: ontwerpend onderzoeken in het sociale domein* (pp. 77-93). Boom Uitgeverij.
- Bereiter, C. (2002). *Education and mind in the knowledge age*. London: Lawrence Erlbaum.
- Dekkers, A., et al. (2005). *Leidraad Communities of Practice in Hoger Onderwijs Community of Practice*. Utrecht: Digitale Universiteit.
- Fischer, G. (2011). Understanding, fostering, and supporting cultures of participation. *Interactions, 18*(3), 42-53.
- Khan, S., & VanWynsberghe, R. (2008, January). Cultivating the under-mined: Cross-case analysis as knowledge mobilization. *Forum: qualitative social research, 9*(1), 34.
- Knorr-Cetina, K. (2007). Culture in global knowledge societies: Knowledge cultures and epistemic cultures. *Interdisciplinary Science Reviews, 32*(4), 361–375. doi: 10.1179/030801807x163571.
- Rombouts, P. (2023). Hoe ontwerp en begeleid je een muzische werkplaats ? Musework. Retrieved from https://www.musework.nl/nl/page/12457/hoe-ontwerp-en-begeleid-je-een-muzische-werkplaats (Accessed: 28 August 2023).
- Ropes, D. C. (2010). *Organizing professional Communities of Practice*. University of Amsterdam.

- Smeenk, W. (2021). *The world tilted, Societal impact and the evolving practice of design – inaugural speech*. Retrieved from https://bit.ly/477gf4E
- Overdiek, A., & Geerts, H. (Eds.) (2023). *Innoveren met Labs 2.0. Ruimte maken voor duurzaamheidstransities*. The Hague University of Applied Sciences.
- Van Turnhout, K., Andriessen, D., & Cremers, P.H.M. (2011). *Handboek ontwerpgericht wetenschappelijk onderzoek: ontwerpend onderzoeken in sociale contexten*. Boom Uitgeverij.
- Van Rosmalen, B. (2016). *The return of the muses; public values in professional practice*. Uitgeverij Ijzer.
- Wenger, E., McDermott, R., & Snyder, W. M. (2002). *Cultivating Communities of Practice. A guide to managing knowledge*. Boston, MA: Harvard Business School Press.
- Wenger, E., Trayner, B., & de Laat, M. (2011). Promoting and assessing value creation in communities and networks: A conceptual framework. *Ruud de Moor Centrum*. doi: 10.1146/annurev.so.16.080190.002251.

Peter Troxler,
Manon Mostert-van der Sar

8. The Open Lab as Boundary Object

DOI: 10.1201/9781003491484

Introduction

It is 2020 when we start a project with four public libraries that want to develop their own makerspaces; places where the general public get access to 3D printers and laser cutters, requiring no former expertise. Under the 'work from home guideline', we coach the librarians online in group sessions and one-on-one. Manon, the head of our lab, shares her decade-long experience and can-do attitude with the librarians to prepare them for their new role as maker coaches. They themselves will have to help visitors to the makerspaces to understand setbacks as stepping stones. They experience this approach themselves, when they struggle designing their first workshop, and she helps them turn mistakes into progress. Says one of the maker coaches, 'She is your biggest fan, even when you're messing up'.

Open labs such as the Makerlab mentioned above are a form of experiential learning environments that are increasingly used beyond being mere facilities to 'make stuff' to connect multi-stakeholders in envisioning, creating, experimenting, learning, and trying out novel responses to diverse societal challenges. An open lab exhibits features that allow the lab to be transformed as needed for co-creation to arena and agora (Jones, 2018) – 'where the lab is a venue for internal research and theory building, the studio a place for internal collaboration and prototyping, the arena a venue for engaging with stakeholders and facilitated (political) dialogue, and the agora is a publicly accessible democratic context' (pp. 21-23). Hence, the open lab has the capability to shift from internal research to structured stakeholder engagement or democratic participation in applied design research. This involves opening the space to committed stakeholders or the public, implementing structured methodologies and dialogue processes, and ensuring a shift from internal development to external, influential engagement.

The contributions in this publication add to the conversation surrounding a list of design dilemmas that ensue from today's societal and environmental challenges. In the perspective we take in this chapter, the open lab serves as the backdrop for the unfolding of some of these dilemmas, not so much in the sense that dilemmas must be resolved, and sides have to be taken, but more in the sense that in a circular rather than linear way dilemmas can be reconciled, and ends can meet. This latter perspective leads us to consider the open lab not just as an experimental learning and innovation environment, but indeed as a boundary object.

Figure 1. Workshop at a library Makerlab (© 2023 Lara Coomans, used with permission).

> Star & Griesemer (1989) define the boundary object as 'an analytic concept of those scientific objects which both inhabit several intersecting social worlds (...) and satisfy the informational requirements of each of them. Boundary objects are objects which are both plastic enough to adapt to local needs and the constraints of the several parties employing them, yet robust enough to maintain a common identity across sites. They are weakly structured in common use, and become strongly structured in individual site use. These objects may be abstract or concrete. They have different meanings in different social worlds but their structure is common enough to more than one world to make them recognizable, a means of translation' (p. 393).

So, the three characteristics of a boundary object are (1) the intersection of social worlds (intersectionality), (2) the structure that is malleable from weak to strong, from common to individual, and abstract to concrete (malleability), and (3) the ability to translate between different social worlds (translation).

In our case, the boundary object we consider is a boundary spanning practice tied to a specific location, namely the open lab – hence the title 'The Open Lab as a Boundary Object'. We approach the question if an open lab as a place that can operate as a boundary object that fosters intersectionality, malleability, and translation. We investigate the boundary spanning practice at Stadslab at Rotterdam University of Applied sciences. The Stadslab itself is loosely modelled after MIT's fab lab concept (Gershenfeld, 2005). Its focus was originally on providing low barrier access to making, sensing, and data, true to Papert's (1980) 'low floor, high ceiling' concept. More recently, capabilities in virtual reality and data science have been developed in the lab. The lab serves primarily as an infrastructure for various academic activities, as a makerspace for students and researchers from applied design research in its core and from a broad range of disciplines, including engineering, healthcare, and the arts (Troxler & Mostert-van der Sar, 2019). Below, we present auto-ethnographic vignettes from our organisational memory (Stein, 1995) and relate them to the three dimensions of the boundary object. An (auto)ethnographic vignette is an ethnographic account providing 'a focused description of a series of events taken to be representative, typical, or emblematic' (Miles & Huberman, 1994, p.81). The vignettes in this chapter are created through multiple encounters of staff with researchers and through reflective workshops with teachers.

Medical 3D Printing

It is May 2013. A medical PhD student from Erasmus Medical Center in Rotterdam shows up at Stadslab. On a thumb drive, he brings a folder with images from the radiology department – CT (Computed Tomography) scans of a fractured hip bone: 'Can you help me turn those 3D images into physical 3D prints?' A quick web search reveals that CT scanners typically output their scans in a standardised DICOM (Digital Imaging and Communications in Medicine) format. And there is free software to view CT scans and another tool to convert DICOM files to STL format, the format that we can send to the 3D printer.

In November of the same year, one of the PhD student's mentors talks about biomedical use of 3D data at a conference on 3D printing. His last slide carries the photo of an Ultimaker Original in a medical lab setting. They had the most sophisticated, high-tech virtual reality visualisation techniques at the hospital, for example for diagnosing embryos. But they said 'you cannot touch them'.

Figure 2. 3D scan of a fractured bone loaded into a 3D modelling software (© 2013, Peter Troxler).

A 3D print can solve that problem, and indeed true size 3D printed models are extensively used in the medical sector for counselling, education, and pre- and perioperative planning.

The medical 3D printing case presents the intersection of making and healthcare, bringing the affordances of a common technology such as 3D printing to the domain of medical operation planning, and translated knowledge from making (the Stadslab) to healthcare (the Erasmus Medical Center).

We like to think that the low-barrier access to 3D printing at the lab contributed to the medical profession's understanding of the affordances of the technology. The lab contributed to the transmission of knowledge between the disciplines – from manufacturing to medical science. It contributed to transforming 3D printing from a manufacturing technology to a tool in medical counselling.

Shared Care

It is still 2013, two retailers visit the lab. They ask, How can they make custom-made stickers for their shop windows? For the lab, this is the first time someone comes up with a practical use for the vinyl cutter. It was initially bought as part of the 'standard inventory' of a fab lab. But people had rarely used it before. Only few people were aware of the affordances of this basic technology – for example, putting out strong, customised messages.

Stickers for promotion and instruction? What works in a retail context can easily work at the lab. At the lab, we developed a set of icons to visualise the capabilities in the lab which we now prominently display on the windows. And, learning from marketing, we put up some slogans and house rules and stuck them to the walls, windows, and doors. Oh, and we learnt something else from the retailers: The lab had carefully prepared instructions for using the vinyl cutter. But for the retailer, they were too convoluted – and partly incomplete. So, together with them, we rewrote the instructions.

Figure 3. 3D printing workstation with the icon for 3D printing on the window behind the computer screen (© 2022 Manon Mostert-van der Sar).

The intersection of marketing and education helped us to understand the affordances of the vinyl cutter and using stickers to communicate specific messages (marketing or educational). Moreover, it led us to collectively translate the instructions from a teaching format to a user's format.

We like to think that the lab is not a closed shop only 'owned' and 'cared for' by the university. Rather, the visitors, too, contribute to the open lab and show ownership and care.

Springboard for Careers

Throughout the years, the lab has been hosted by stewards, a multidisciplinary group of students that study at the Rotterdam University of Applied Sciences. This group of students makes their own work schedule, keeps the machines alive, and helps others to use the lab. Despite regular changes due to new students starting internships or graduating, the team remains tightly knit, often found collaborating even outside their designated work hours. The students, previously scattered across separate locations, would not have crossed paths without the lab. New friendships are often born within this environment.

Friendships can evolve from the safe zone of the lab into real-life ventures. For instance, a group of former stewards engineered a lamp powered by the earth's energy from plants. Similarly, another group of stewards, started a pop-up store featuring design ware. Many stewards transitioned into primary or secondary education, combining teaching and technology, their experiences in the lab serving as a springboard for their careers.

The lab not only acted as a place for interdisciplinary encounters, it also operated at the intersection of student and young professional. Some stewards used the informal setting of the lab to develop more stringent relations as in starting a venture together. For them, the lab facilitated the translation from employee in a side job to entrepreneur.

We like to think that the lab for these young people transmutes from a safe zone of experimentation to a place for real-life experience. Their reciprocated encounter of diverse expertise fuses into ventures that thrive on innovation. Nowadays, not only students from higher professional education are using the lab but also students from secondary vocational education and secondary education, making it even more likely to meet diverse expertise.

Figure 4. Former stewards posing as a start-up company for the Dutch Design Awards (© 2017 Ruud Balk, used with permission).

Transformative Learning

Fast forward to 2022. Researchers and teachers from the lab are working with public libraries that want to establish their own makerspaces. TU Delft students work the challenge to design interactive concepts that support the development of the library makerspaces. Where possible, libraries put those concepts into practice themselves. Often, the lab experts give them a hand in doing so. One of the concepts is an interactive table to discuss issues of urbanism with the public. This concept, too, is a bit too challenging to be developed in a starting makerspace. So, Stadslab Rotterdam takes charge of that development. Two young design teachers adopt the project and build a first prototype. Through many iterations, they gradually improved the interactive table over the course of a year until they are satisfied that the concept works – and they are confident to give a workshop themselves to the maker coaches from the libraries during which the coaches can build the interactive table themselves – under careful instruction.

The workshop went extraordinarily well, and ten public libraries in the Netherlands now own an interactive table and are collectively developing a practice around it. This is a fine result in itself. For the two design teachers, the experience was far bigger. They learnt themselves what it means to build a prototype, improve it, and develop it until it can be successfully shared with others. Of course, they knew abstractly about prototyping, and indeed they were teaching 'prototyping' to students. However, through their own experience, they developed a deeper personal and practical understanding of prototyping. One student said, 'I also find that it makes the dialogue with students about making and prototyping a lot more tangible and relevant for both parties.'

In this vignette, the intersection was – rather abstract – between the cognitive and the embodied. The design teachers were enabled to take what they knew from theory as prototyping and experience prototyping in their own practice. In doing so, they translated cognitive knowledge to embodied knowledge, from textbook to (bodily) experience.

Figure 5. Demonstration of the interactive table (© 2022 Casper ten Burgh, used with permission).

We like to think that the lab created a transformative and long-lasting experience for the two teachers. In experiencing prototyping work, their generic, yet abstract, knowledge of prototyping got transformed into an embodied practice. However, as prototyping is a fundamental ingredient of designing, this transformation of knowing at the same time renders knowledge more robust.

Citizen Science

October 2023. It is the 'CityClimate meets CreativeCoding' festival in Hamburg, Germany. Part of the festival is a workshop on climate change awareness, a special workshop as it is not centred mainly around post-its and flip overs. Rather, participants are given a robotic platform – the WijkBot (Dutch portmanteau for neighbourhood (wijk) and robot), a low trolley or cart that can be remotely controlled and driven through streets and alleys. These trolleys serve as a platform for building all sorts of machines, known examples are package delivery, cleaning, or policing carts. In the Hamburg workshop, participants built carts that would broadcast podcasts, carry a catapult for launching and catching climate change-related questions, and a robot collecting and immediately redistributing funds.

The WijkBots were taken to public spaces in the city of Hamburg for interactions with passers-by. The bots appear to be appealing civic prototypes (Jaskiewicz, 2022) that lower the barriers for interaction between design researchers and the general public. The WijkBot platform allows everyone to actively contribute to the shaping of an unpredictable technological future, emphasising the collective power in influencing innovation. This work underscores the essential role of applied design research in empowering individuals to engage with and impact technological progress, emphasising the importance of a shared voice in steering the course of innovation. The lab makes it possible to not only create, but also test and reflect on the prototypes.

Figure 6. WijkBot workshop with the cart sitting on a table while mounting the prototype (© 2023 Tomasz Jaskiewicz, used with permission).

This vignette highlighted the intersection of academia and civil society, giving the latter an opportunity to play with near future technology the former is developing. The WijkBot itself as a prototyping platform performed as a generic tool that could be rendered specific by the participants. So, abstract technology was translated to practical, while still speculative, applications.

We like to think that civic prototyping entails accessible creative prototyping, offering individuals the means to shape the role of technology in fostering what's termed as 'democracy in the small' (Jaskiewicz, 2022), hence rendering the fixed roles of (design) researcher and test subject completely fluid.

Theorising the Open Lab as Boundary Object

An open lab, distinct from other types of labs, cultivates a collaborative environment that promotes interdisciplinary interactions and embraces individuals from diverse backgrounds and expertise. While many fab labs and makerspaces align with these criteria, their organisational structure or close-knit community often creates spaces where only like-minded individuals converge. However, the fundamental difference lies in the open lab's emphasis on fostering inclusivity, encouraging a broader range of participants to engage and contribute, ensuring that diverse perspectives and ideas intersect and foster innovation beyond the confines of a specific community or niche. So, an open lab, such as Stadslab Rotterdam, exhibits the features of a boundary object – (1) it inhabits intersectional worlds, (2) its consistency is malleable, and (3) its structure is common enough to be a means of translation.

For applied design research, the open lab, as boundary object, arena, and agora, provides a locale for engagement in the practical aspects of design research. The dilemmas in design research that stem from the wicked problems of diverse societal challenges are not to be resolved, but to be addressed, and differences to be navigated, and to be transformed to novel solutions that would not have been possible without the recognition of these dilemmas and differences, a process known as boundary crossing (Suchman, 1994; see also Troxler, 2022).

Practising the Lab as Boundary Object

To make the lab useful as arena and agora – and operating from an educational context – it is important to come up with new ways to organise the lab. Stadslab Rotterdam, for example, is always open for every student and teacher 5 days a week, but also open to the public for two moments a week, every week. This makes it possible for different stakeholders to visit the lab and interact with both technology as makers and so engage in collaborative innovation, fostering a dynamic environment where diverse perspectives converge, echoing the essence of (1) inhabiting intersectional worlds.

Furthermore, co-creative educational methods can foster an environment where diverse disciplines naturally converge. Within Stadslab Rotterdam, a diverse range of offerings has been developed, allowing the earning of Edubadges – digital certificates displaying various skills. This offering operates within an open structure, devoid of credit points, irrespective of study background or academic year, and is both demand and supply driven. These modules bring together students and educators from various backgrounds, enabling them to learn and practise a particular skill collaboratively, across different proficiency levels. Particularly, teaching in such an environment requires a new attitude which is designerly and driven by a teacher's own curiosity (Mostert–van der Sar, 2019). These Edubadges also create an open space for educational development making it possible to change and adapt over time, corresponding with the essence of (2) consistency is malleable, which indicates that the nature or characteristics of the open lab can change or adapt over time. It's not rigid or fixed in its structure or purpose. Instead, it remains flexible, allowing it to evolve based on the needs, interests, and input of the participants and stakeholders involved.

Another way to stimulate encounters is by facilitating low-threshold connections among lab users. Stewards are trained to create links between diverse groups within the lab by pointing out each other's work or actively encouraging them to assist one another. For example, when users must wait for a machine, stewards encourage them to help each other, reducing waiting times. This has already resulted in several serendipitous encounters that led to unusual and innovative collaborations, such as the joint development of smart clothing by a computer science and fashion student, neither of whom had previously expressed interest in each other's fields.

Planned structured educational modules can also transform the open lab into an arena with diverse stakeholders. For instance, through the implementation of a time-boxed program, where students work continuously for several days on a smart prototype for a complex problem, integrating making, testing, and reflection as crucial components. Whenever feasible, stakeholders are involved before, during, or after the project. By assembling multidisciplinary teams, complex dilemmas can be openly addressed. This echoes with (3) a common structure that enables communication and collaboration among its diverse participants. Despite the differences in backgrounds and expertise, there are shared elements or principles that facilitate understanding and cooperation, allowing ideas to be exchanged and translated across different perspectives.

Discussion

With these different vignettes from the lab, we intended to demonstrate that the lab is like a meeting point where different ideas come together and make new things happen. The experiences shared here show how the lab helped turn knowledge and technology into real changes that affected society.

Intersectionality

Where different worlds meet. The lab embodies a meeting ground for diverse realms, where the convergence of various disciplines and perspectives sparks innovation and learning. In two ways specifically, this meeting of worlds renders the lab a boundary object: First, as a space where the affordances of technologies come into effect, where they transcend mere functionality to become transformative tools shaping medical, educational, and societal landscapes. Second, more than a physical space, the lab is an incubator where knowledge gets transformed, shifting from abstract understanding to practical, hands-on experience.

Malleability

Where abstract affordances (of technologies) are turned into practical effect. The open lab serves as a dynamic platform where insights are shared and discoveries are made. These discoveries lead to very practical applications of the promises of technology – not only in the utilitarian realm of application (what problem can we solve with a specific technology?), but also in the entrepreneurial and speculative realms (what are commercial prospects, and what are societal challenges a certain technology relates to?).

Translation

Where knowledge gets transformed. Ultimately, the lab embodies a space of transformation where the fusion of technology, knowledge, and diverse perspectives shapes our collective understanding and applications. It's an environment where innovation and creativity flourish, paving the way for a more inclusive, technologically empowered, and knowledge-rich society.

The lab plays a key role for encounters and as place of innovation:

- It allows for experimenting, sharing of insights and knowledge in general;
- It is a transdisciplinary place for development and learning;

- It is inclusively sociable, involves social and cultural life, and becomes a public understanding of science.

Through the vignettes shared here – from medical imaging applications to signage production and interactive concept development – the lab appeared as a catalyst for evolution, not just in the academic sense but in the broader context of social and cultural integration.

Conclusion

Labs play a key role for encounters and as places of innovation – they allow for experimenting, sharing of insights and knowledge in general and education and training more specifically in a higher education context (Mostert–van der Sar et al., 2013).

In a higher education context, the lab (in its narrow sense as defined by Jones (2018)) primarily works as a place of work for education and research. But in the lab, we often hear students say: 'I have to go back to school'; even though they are already there, they don't experience it that way. That points to more ways to look at and understand the lab, which should be the subject of further research. On the one hand, the open lab as a third place (Oldenburg, 1989) is inclusively sociable as a hangout. On the other hand, one might further want to explore the lab's relation to the third mission of universities (Etzkowitz & Leydesdorff, 2000) – under the condition that the third mission moves from utilitarian definitions (intellectual property, spin-offs, policymaking), to involvement in social and cultural life and public understanding of science (Laredo, 2007).

The lab as a boundary object manifests not as belonging to one or the other end of a dilemma such as open or closed, disciplinary or transdisciplinary, transformative or abiding. Rather, the lab forms the arena where dilemmas can unfold, and where contradictions are welcomed, accommodated for, and serve as a fertile ground for intersectional encounters, allow for general, common structures to be rendered more specific and individual, and facilitate the translation of knowledge and meanings between different worlds.

Bibliography

- Etzkowitz, H., & Leydesdorff, L. (2000). The dynamics of innovation: From National Systems and 'Mode 2' to a Triple Helix of university–industry–government relations. *Research Policy, 29*(2), 109–123. https://doi.org/10.1016/S0048-7333(99)00055-4
- Gershenfeld, N. (2005). *Fab: The coming revolution on your desktop, from personal computers to personal fabrication.* Basic Books.
- Jaskiewicz, T. (2022). *Civic Prototyping: A Creative Encounter Between Design Prototypes and Engaged Citizens.* Hogeschool Rotterdam Uitgeverij.
- Jones, P. (2018). Contexts of Co-creation: Designing with System Stakeholders. In P. Jones & K. Kijima (Eds.), *Systemic Design: Theory, Methods, and Practice* (pp. 3–52). Springer Japan. https://doi.org/10.1007/978-4-431-55639-8_1
- Laredo, P. (2007, March 5). Toward a third mission for universities—UNESCO Digital Library. *Globalizing Knowledge: European and North American Regions and Policies Addressing the Priority Issues of Other UNESCO Regions, Paris.* https://unesdoc.unesco.org/ark:/48223/pf0000157815
- Mostert-van der Sar, M. (2019). *Hey Teacher, Find Your Inner Designer.* Boom Uitgevers. http://boomhogeronderwijs.nl/heyteacher
- Mostert-van der Sar, M., Mulder, I. J., Remijn, L., & Troxler, P. (2013). Fablabs in design education. *Proceedings of E&PDE 2013, International Conference on Engineering and Product Design Education,* 629–634. Retrieved from file:///C:/Users/PoeHG/Downloads/DS76_101.pdf
- Papert, S. (1980). *Mindstorms: Children, computers, and powerful ideas.* Basic Books, Inc.
- Star, S. L., & Griesemer, J. R. (1989). Institutional Ecology, 'Translations' and Boundary Objects: Amateurs and Professionals in Berkeley's Museum of Vertebrate Zoology, 1907-39. *Social Studies of Science, 19*(3), 387–420.
- Stein, E. W. (1995). Organization memory: Review of concepts and recommendations for management. *International Journal of Information Management, 15*(1), 17–32. https://doi.org/10.1016/0268-4012(94)00003-C

- Suchman, L. (1994). Working relations of technology production and use. *Computer Supported Cooperative Work, 2*(1), 21–39. https://doi.org/10.1007/BF00749282
- Troxler, P. (2022). Dance? Dance! The contribution of practice-driven design research to the ballet of disciplines. In P. Joore, G. Stompff, & J. van den Eijnde (Eds.), *Applied Design Research: A Mosaic of 22 Examples, Experiences and Interpretations Focussing on Bridging the Gap between Practice and Academics* (pp. 63–71). CRC Press. https://doi.org/10.1201/9781003265924
- Troxler, P., & Mostert-van der Sar, M. (2019). *Seven Years of Plenty? / Zeven jaar rijkdom?* (2nd ed.). Kenniscentrum Creating 010, Hogeschool Rotterdam. https://hbo-kennisbank.nl/details/sharekit_hr:oai:surfsharekit.nl:aabd229b-ca20-4771-9ee5-a86fe6f1fad5

Part 3: Materialisation of the Living Lab

Tiwánee van der Horst,
Anja Overdiek, Maaike Harbers

9. Bridging Multi-Stakeholder Dialogue about AI Systems in the Lab: How Virtual Can We Go?

DOI: 10.1201/9781003491484

Introduction

Emerging technologies like artificial intelligence and augmented/virtual reality come with many ethical and societal questions. In order to develop applications based on these technologies responsibly, businesses, governments and knowledge institutions are increasingly working together, and sometimes also involve citizens. To tackle the challenge of the 'responsible digital transition', Living Labs are often chosen as a methodology for triple- or quadruple-helix co-creation and experiment. One example are the Dutch ELSA labs introduced in 2019 by the Dutch AI Coalition as open labs tackling ethical, legal and societal aspects in the development and application of artificial intelligence (Van Veenstra et al., 2021). Societal aspects of responsible digitalization also span across the ecological and societal underpinnings and repercussions of technologies. AI systems (Sartori & Theodorou, 2022), for example, rely on an enormous amount of energy, scarce raw materials, consumer manipulation and unethical labour arrangements to train algorithms (Crawford, 2021). These systemic issues are complex and intangible, in part because of their interrelatedness with many different stakeholders and invisible power dynamics that are at play. The dialogue necessary between lab stakeholders thus increases in complexity when systemic issues are introduced. Nevertheless, the AI practice is multi-disciplinary by nature and responsible AI development will benefit from being aligned within dialogues about policymaking, regulatory processes, and design decisions.

Co-creation and experiments in Living Labs can be used within different constellations of stakeholders, in a more closed or open collaboration (Jones, 2018). However, before experiments can be conducted, the group of stakeholders needs to align on a common specific task. For example, when companies from the business services sector, governments and researchers find each other to address the challenge of developing 'responsible applied Artificial Intelligence' (Harbers & Overdiek, 2022) they still need to agree on a more precise application field for their design question related to real-life uses. During the co-creation sessions leading up to finding this common task, bridging the different perspectives, interests and languages of the stakeholders is a fundamental problem which designers or design researchers intend to tackle (Aguirre et al., 2018; Design Council, 2020). They need to facilitate boundary spanning.

Boundary Objects as Boundary Spanners in Living Labs

Design and organisational research scholars argue that boundary objects have proven to be an effective method to communicate systemic properties and bridge stakeholder perspectives in collaborative processes (Carlile, 2004; Cooney et al., 2016; Star & Griesemer, 1989). Boundary objects have boundary-spanning capacities which are especially relevant within Living Lab settings in which multiple stakeholders and contexts need to align towards a common task.

Boundary objects can be emergent or designed. Emergent boundary objects are existing objects that become boundary objects as they are used by different stakeholders to bridge perspectives (Star, 2010). Designed boundary objects are objects that are specifically designed to perform a selected boundary-spanning capacity for a specific context in which this boundary spanning is needed. Boundary objects can also be physical or virtual. In this chapter we elaborate on experiments done with a designed physical boundary object (PBO) in physical and virtual space. This PBO was designed for a specific context of use that asks for boundary spanning between social worlds of multiple stakeholders within the AI ethics context.

Table 1. Types of boundary objects with examples.

Types of Boundary Objects	Emergent	Designed
Physical Boundary Object (PBO)	Sketch drawing, clay	Model, map, physical game
Virtual Boundary Object (VBO)	Excel sheet, 3D modelling tools	Micro-enabled interaction, online game, simulation

Our research into the workings of a designed PBO, to bridge perceptions and reframe dialogue to a more systemic level, touches the polarity of a Living Lab being a physical or a virtual place. Table 1 shows some examples of the different types of boundary objects. Our designed PBO focuses on the multi-stakeholder co-creation phase taking place even before any 'real-life' experiment can be conducted in the lab. Dietrich et al. (2021) argue that this is a phase where co-design methodology can contribute to existing Living Lab practice. This is also the phase which frames the design space for the later concepting and testing phase in a lab, often described as a 'dialogue' phase (Yasuoka et al., 2018).

In university Living Labs, applied research and education are fostered by using the campus and its facilities to develop and test real-time solutions in digital transition processes, thereby offering opportunities to students, teachers, and users of AI to learn in daily life (Van Geenhuizen, 2018). On-campus experimentation largely consists of integrating inventions into existing social, building, urban, technological, and/or infrastructural fabrics. As shown in Figure 1 by Yasuoka et al. (2018), the dialogue preceding this challenge definition is crucial. The goal of such dialogue is to find a common ground. In a multi-stakeholder dialogue, multiple power structures are at play, varying between people with different backgrounds in terms of ideology, politics, culture, (technological) accessibility, language, knowledge, work, and life experience, etc.

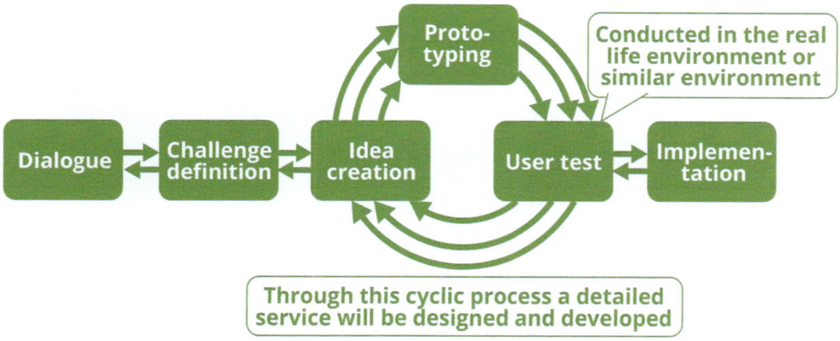

Figure 1. Living Lab process according to Yasuoka et al., 2018 (page 133).

Physical and Virtual Living Labs

During the COVID-19 pandemic, Living Lab activity was moved online, forcing practitioners to develop new lab methodologies (Dhawan, 2020). Advantages of virtual labs like the inclusion of elderly or the connection with a broader network of users (Haukipuro & Väinämö, 2019; Thomas et al., 2022) are increasingly stressed. Mačiulienė & Skaržauskienė (2020) argue that particularly in the co-creation of public spaces, information and communication technology (ICT) can be a powerful tool to achieve inclusiveness, to contribute to accessibility of information, and to enable new forms of collaboration. However, new boundaries arise when dialogue needs to happen in the digital realm (Choi & Cho, 2019; Karl et al., 2022; Morrison-Smith & Ruiz, 2020).

The research in this chapter centres around a PBO called 'Who pulls the strings of AI'. With this PBO, we aimed to bridge different perspectives and invite dialogue about power dynamics behind AI systems in a physical multi-stakeholder lab setting. The physical facilitation with the PBO showed promising results in bringing together 10-20 stakeholders around the issue of power and AI per session. However, we found that the PBO is limited in reaching a larger group of people, because of its physicality. A common way to increase scale and flexibility in collaborations has been to host them in a virtual or in a hybrid form. But how virtual could we go with this particular lab intervention? This question touches the broader polarity related to Living Labs: when and how are hybrid or online environments favourable, and when is it better to stick to physical interaction?

In this chapter we will take a closer look at both physical and virtual facilitation of our PBO, to find out how and if this PBO can be used in online or hybrid sessions. We analyse the dialogues and qualitatively look at how the physical and virtual sessions lead to finding a common task. By analysing and discussing this case, we formulate conclusions about the use of this particular PBO and some hypotheses about using boundary objects in hybrid or online lab environments. In the following sections we will describe (1) the design of a PBO facilitating a systemic dialogue about power relations behind AI in a lab, (2) an analysis of physical and virtual facilitation of the PBO, (3) conclusions about opportunities and challenges in using a BO in physical and virtual Living Lab environments, and (4) suggestions for future design of a virtual boundary object (VBO).

The research question we address in this chapter is:

- *How do virtual and physical facilitation of a designed physical boundary object contribute to bridging perceptions and reframing multi-stakeholder dialogue in the Living Lab process?*

Methodology

In a previous paper, 'Designing a Physical Boundary Object to Invite Dialogue about Power Relations Behind AI' (Van der Horst et al., 2024), we described the concept and design principles that led to our PBO (see Figure 2). In this chapter we will not elaborate on these design principles, rather we will focus on the use and effectiveness of the PBO as a design intervention in a virtual and physical setting.

Use case

At Rotterdam University of Applied Sciences, the research group for Responsible Applied Artificial InTelligence (RAAIT) deals with AI ethics in which many challenges in boundaryspanning arise. On the input side, boundary spanning needs to happen between multiple disciplines from software engineers, ethicists, policymakers, communication managers, designers, developers, students, researchers, and more. These stakeholders all need to unite over a common challenge before moving towards idea creation (Figure 1). To test the PBO in the dialogue phase, we used the context of the Mozilla Festival (short: MozFest). MozFest is a gathering of a global community of technologists, designers, artists, hackers, start-ups, NGOs, and researchers that engages with internet health and technology ethics. This festival is not a Living Lab; however, this context for dialogues is similar to that of the Living Lab context. This is because the participants of the workshops wish to have a dialogue about AI ethics and come together to share ideas. MozFest is known for its power to create unexpected partnerships to redesign our online environment, with a focus on trustworthy AI.

During the COVID-19 pandemic, MozFest had gained experience in hosting the festival online. The online festival allowed for internet health stakeholders from across the globe to meet and engage in dialogue while being in different time zones. However, there was still a need for people to meet in real life. When the world opened again in 2023, the MozFest organisation saw relevance in continuing the online festival and also providing local and situated regional programs. Like that, the physical festival also supports local initiatives to grow towards the next virtual MozFest. MozHouse 2023 took place in a physical venue in Amsterdam. Here, multiple workshops, (art) installations, presentations, and panel discussions were held in parallel in the same building during 2 days. We were invited to bring the PBO 'Who pulls the strings of AI?' to the physical festival after having done the online workshop. These two different contexts within a global community of technology ethicists from diverse backgrounds allowed us to experiment with the PBO in different contexts, namely the online version of the MozFest (March 2023) and the physical MozHouse (June 2023) in Amsterdam.

Figure 2. The physical boundary object (PBO) 'Who pulls the strings of AI?'.

Data collection and analysis

To answer the research question, we performed interventions with the PBO in a physical and in a virtual workshop context, collecting data by participatory observation and transcription of the resulting dialogues. We will analyse the observations and the dialogues that resulted from the two workshop iterations by comparing the two datasets, extracting leading themes and interpreting them (Alhojailan & Ibrahim, 2012; Miles & Huberman, 1994). This method is compatible when qualitative data is collected and compared in different phases and settings during a project. We will also discuss these results against the polarity of facilitating multi-stakeholder dialogue in physical and in virtual settings in multi-stakeholder labs.

The contributing stakeholders of the empirical testing in a physical (Figure 3) and digital (Figure 6) workshop at MozFest were diverse participants who chose the workshop themselves. In both cases we experimented with the PBO in a setting where the participants met each other for the first time. The aim was to invite dialogue about power relations behind AI. In both cases, the discussion was held in 1 hour. To give each participating stakeholder the opportunity to speak, listen, and respond, the group was restricted to a maximum of 15 people. In effect, 3 people participated in the online, and 13 people participated in the physical workshop. The workshops were facilitated by the first author who also designed the PBO. The second author made verbatim notes of the dialogues. Prior to the workshop, we asked permission of the participants to use the data for research purposes. Data analysis was done by all three authors.

Figure 3. Facilitator explaining the use of the PBO to participants at MozHouse, June 2023.

Use of the physical boundary object

In the following section, we describe how the PBO was used in the two workshops. For test 1, physical lab, the PBO was placed on a rectangular table of 100 cm height, and participants stood around the table. This enabled all participants to see the object and easily reach the interactive parts. For test 2, digital workshop via ZOOM, the PBO was placed on a table and a camera connected to the festival's collaboration platform was focused on the object. A digital twin of the board (Figure 4) was designed in Miro and facilitated online, and the second author replicated the online 'moves' of the participants in the physical setting. Like that, the participants could witness the effect of their moves on the object via the film footage. In both workshops the PBO was introduced, and then the participants were asked to go through the same process questions (Van der Horst et al., 2023):

Figure 4. The digital twin of the board showing the quadruple-helix distribution where the pawns could be placed in the grid.

1. To what extent is your voice being heard in conversations about AI? Choose pawn weight: none, small, medium, large.
2. In what field do you position your daily role?
   Choose a quarter of the board: government, knowledge institution, industry, user3
3. How much influence on the development of AI do you have in this role?
   Centre of the board is no influence, outer edge of the board is large influence.

All participants place their pawns with chosen weights, and as a consequence, the balancing board, hanging from ropes, tilts towards the heaviest side of the board, illustrating the power (im) balances within the group of participants. The tilted board affects the movement of the puppet hanging below the board. Thus, the hanging board, the pawns, and the puppet together physicalize a metaphor for power 'Who pulls the strings'. After placement of the pawns on the board, participants are asked to explain why they choose their weight and positioning on the board. When everyone has explained their perspective, the facilitator initiates a reflection on the balance of the board. Participants have the option to adjust their position or react to other participants' position.

Even though we asked the same questions and used a similar facilitation approach, we acknowledge that the session results are not one-on-one comparable. As such, we qualitatively look at what conversation mechanisms occur in the physical and virtual sessions. We look at when and how perspectives are successfully bridged or when and how obstructions arise. In doing so, we can determine what mechanisms lead to a reframing of the multi-stakeholder dialogue.

Results

In the following paragraphs, we summarise key insights of the physical and virtual session with the PBO. This is enriched by quotes of the different dialogues.

Physical workshop

Engagement

The participants were from various backgrounds such as journalism, NGOs, design research, IT developer, student, and users of AI. Representatives from governments were missing. The participants did not need a lot of explanation to be able to use the boundary object as it was intended.

In the physical workshop, participants faced the PBO and each other. During the individual round, they made eye contact and talked one after the other. They also watched closely the effect their action (putting their pawn on the board) had on the PBO (how the balancing board tipped). During the follow-up dialogue, they often had their eyes on the board and kept referring to it, to clarify their statements. Although there were 13 participants, almost everybody contributed to the dialogue within the timeframe of one hour. The conversation did not fall silent at any moment and the facilitator only interfered twice: when a question about the PBO came up and in order to lead from the dialogue phase to the evaluation phase of the session. Thus, we can say there was a high level of engagement between all participants.

Figure 5. The board at MozFest after a part of the pawns was placed.

Interaction and reflection

Stakeholders pointed towards other stakeholders' positions on the board when discussing their influence on the AI system and realising their interrelatedness. For example, a participant from an international NGO places her pawn with a medium weight in between the government and user corner and halfway between centre and the edge of the board. Pointing to a user pawn, she explains: *'We are really influential in policy debates within the EU, but we have less influence on the public debate'.*

In addition, participants started to see where stakeholders that were not present in the workshop needed to be positioned on the board (see the empty corner on the top, in Figure 5) . They realised that some stakeholders, such as governments and Big Tech, were not present, but concluded they needed to be represented as 'heavy' pawns on the board (and thus seen as part of the system). *'It is remarkable to see that the government is underrepresented at this event. It would be good to invite them and include them in the Mozilla Community'*, concluded a participant. Another participant made the board tip towards the industry. She explained that she had chosen a larger weight because she felt that it was needed relative to the choice of weight of the other participants. If she had been first to choose, she would have chosen a lighter weight. This demonstrates how the participants relate to each other during the use of the board with the particular dynamics of this PBO.

Potential of PBO for reframing and alignment of multi-stakeholder dialogue

In the physical workshop, the interrelation of the influences of the different stakeholders was discussed. The conversation between the participants moved between power constructs on a global level as well as on a community and individual level. This scoping occurred naturally during the dialogue, as participants voiced their perspectives. Participants were able to step into each other's shoes and respond to others' positions even though the positions were significantly different from each other. One participant introduced us with a relatively unheard perspective, she said: *'With my regional NGO office in the Middle East and North Africa, I feel more like a user, because the region is waiting for the EU to make policy decisions. I place my pawn at the edge of the board, with a medium weight, because I think users could move the discussion because they are many'.*

Other participants responded to this actively. It was clear that the participants became aware of their level of influence as European stakeholders. The PBO functioned as a common ground for the participants to refer to, placing them all in relationship to each other. At the end of the physical session, the participants developed the idea of suggesting the instigation of an online open-discussion platform on AI to the broader community of the conference. Thus, a dialogue of approximately 50 minutes with stakeholders from industry, NGOs, knowledge institutions, and users led to a concrete and shared approach related to systemic aspects of AI.

Online workshop

The online facilitation of the PBO went quite differently. In the virtual session there were three participants: one from the NL (A), one from South Africa (B), and one from Spain (C). After explanation of the interaction with the boundary object, the participants were still unsure of how to use the digital twin on the Miro board they were provided.

Engagement

During the entire session, guidance was needed from the facilitator to help them place their pawn. Participants expressed and asked for clarification for what was meant by the process questions asked by the facilitator. It took about 5 minutes before the first two participants decided on a placement for their pawns. Afterwards, there was a discussion on whether this placement was correct. This could mean that the digital aspect of the assignment created an atmosphere in which a mistake was not welcome. Another possibility is that participants could not try to put their pawn on the PBO themselves and directly experience the results, like they did in the physical workshop. This possibly increased an action threshold and decreased a sense of agency over the placement of their pawn and its subsequent effect on the board. It can be concluded that the online engagement with the topic was hesitant and needed active facilitation.

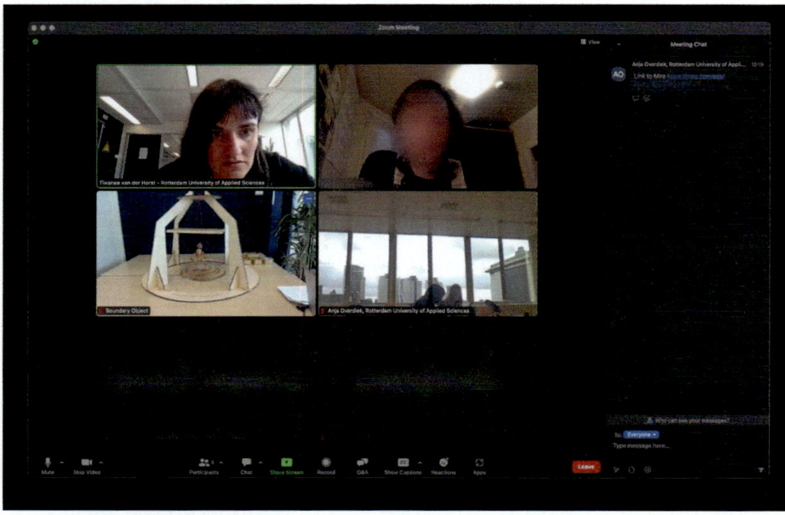

Figure 6. Screenshot after the start of the online workshop with the facilitator (above, left), one participant (above, right), the filmed PBO (middle, left) and the second participant (below right), the third participant arrived after the screenshot was made. Screenshot made by Philo van Kemenade.

Interaction and reflection

The three participants chose and placed their pawn without explanation. As they could only see each other's heads on the screen, it was difficult for them to relate to each other's action (see Figure 6). They placed their pawns simultaneously on the digital twin (see Figure 7), because the digital twin of the board in Miro allowed them all to give their input at the same time. After placing their pawn, they did not immediately experience the change their pawn was making on the balance of the board. The facilitator then had one-on-one dialogues with the participants about their choices, and the participants didn't react to each other. During these dialogues, other participants would sometimes add or move a pawn on the Miro board. Thus the activities remained very unrelated.

On top of that, participants did not relate their reflections on power relations in AI to the feedback of the PBO (by, e.g., pointing to it or mentioning the change in the object they could see on video). Neither did they relate their own reflection to the position of pawns of other participants. Like that, reflection remained very subjective with little interaction with other participants.

One example

Participant A mentions which colour his pawns are. He says he is positioned in both knowledge institutions and industry, as he has several job positions. *'I am busy with integrating AI for users of media archives and I do AI consultancy within the industry. People know I have knowledge about AI'.'* He asks for clarification on whether weight and position of a pawn make a difference in meaning. Participants A and B discuss with each other how the board works. B says, *'I think it means the same thing'.* B continues talking and starts explaining the power mechanisms at play in the Eastern Cape area explaining the position of his own pawn.

This illustrates how the participants were easily inclined to talk past each other, which resulted more in subsequent monologues than in a dialogue. Next to that, the PBO was only mentioned once in the whole discussion, by the facilitator. That was when the board tipped significantly and she asked the participant who had last placed a pawn: *'How do you feel about the balance now?'* After which the participants started looking at the video of the PBO for the first time. It seems like the delay and physical distance between 'placing your pawn' virtually and the physical placement of the pawn prevented the participant from really interacting with the board and each other.

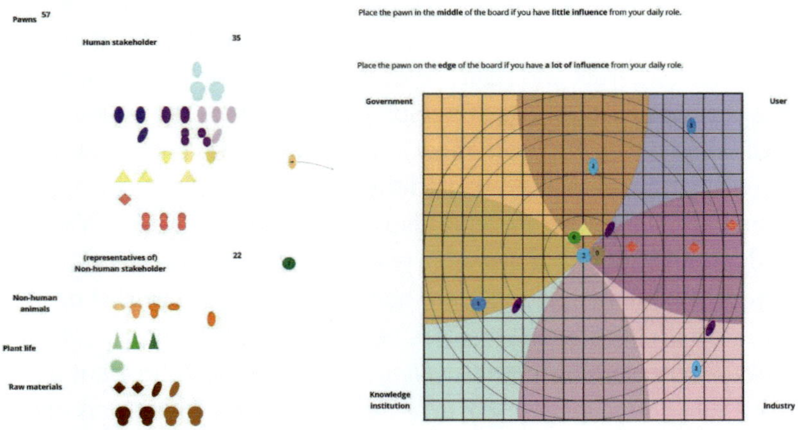

Figure 7. Screenshot of the Miro board: Digital representation of the pawns (left) and end result of the digital twin of the PBO (right).

Potential of PBO for reframing and alignment of multi-stakeholder dialogue

In the online setting, the dialogue moved away from the influence of different stakeholders towards a critical reflection on the level of abstraction of the PBO design. This 'meta-conversation' influenced the dialogue in such a way that it did not flow naturally. The participants in the online workshop were mainly concerned with individually choosing the 'right' pawn's weight and placement. They ended in splitting up their individual influence between different stakeholder pawns, i.e., that of the knowledge institution, that of the user, and that of the government representing different situations of their daily practice. To illustrate, B put several pawns representing multiple positions within industry, user, government, and knowledge institutions. It is unclear if B works in all of these or if he is trying to map out an overview of his field of stakeholders, as he does not clarify this in the dialogue. Participant A starts to do the same, *'B inspired me to break up my pawn into three pawns because I wear more hats'.* He chooses a pawn for industry, government, and users. *'I adjust my industry pawn to the middle just a bit because of how much code is used in R&D. As a user I choose weight 1 and position it close to the centre, because I don't have influence on AI algorithms'.*

We can say that the dialogue was predominantly about the 'right' use of the PBO. As a result, the dialogue about power did not go into depth and remained general. This and the fact that they didn't interact and relate to each other's positions resulted in a very abstract sharing of personal experiences. The online facilitation of the PBO somehow individualised and complicated the dialogue and thus prevented the multi-stakeholder exchange from happening. The dialogue also quickly changed from scale and scope. Suddenly a new idea was introduced: Participant C enters the online meeting room halfway through the workshop and starts reflecting on what has been placed on the board . *'It looks like a government that is falling flat on its face'.* Participant B asks, *'What is the relationship in terms of tipping the board? It is not how I relate to the dynamics'.* Participant C asks if there is danger in abstracting systemic mechanisms in such a model; *'I am thinking of the impact that is not visible now, for example living in Spain and the energy someone is using in Manila'.* At the end of the online workshop, the participants said they liked how the PBO had invited them to reflect on their own influence. The PBO had helped them with that, but didn't align their individual reflections to a shared approach. The participants expressed that they felt limited because their individual contexts were different from each other. They also reflected on the process by recognizing that they were struggling to have a dialogue because the PBO was designed to be used in a physical setting rather than an online setting.

Discussion

The aim of this chapter was to find out how virtual and physical facilitation of a designed boundary object contribute to bridging perceptions and reframing multi-stakeholder dialogue in the Living Lab process. It is not surprising that the virtual workshop was less successful than the physical workshop, especially since the design of the BO was intended for physical use. However, the virtual workshop allowed us to identify where the PBO succeeded in boundaryspanning in the physical setting, and where the PBO failed in doing so in the virtual setting. The virtual workshop confirmed the value of direct feedback and individual interaction with a BO in physical space to start a more systemic reflection process. We can say that, if one wants to use a BO in an online setting, a BO needs to be designed specifically for that. The workshop with the PBO informed us to formulate a design suggestion for a virtual BO (VBO). We will address this first. In the following section, we will reflect on the specific challenges we encountered in terms of engagement, trust, and interaction in the virtual workshop. We will discuss this against existing literature in online (lab) collaboration.

Translation of a Boundary-Spanning Object from 3D to 2D

In the physical and virtual workshop, we used the PBO as a tool to bring the topic of power dynamics behind AI into the dialogue. However, the PBO was designed specifically to be used in a physical setting. The designed physical action of choosing a weight, attaching it to the pawn magnetically, and placing it on the board results in a movement of the board and a physical and visual feedback to all participants at the same time. Standing around the board allows people to make adjustments for themselves, walk around the board, and point to things. However, in the virtual workshop, participant actions are mediated only via the visual screen and voice. There are several barriers influencing action and interaction. Within these barriers a translation needs to be made: between participants, facilitator, facilitator assistant, and boundary object. This means that, where in real life the participant can directly interact with the BO, in digital space there are many barriers that potentially affect how meaning is translated. The barriers then become the following: Boundary Object > Facilitator assistant > Facilitator > Screen & voice > Participant > Screen & voice > Other participants.

In virtual space, the dialogue becomes linear and serial. One participant after the other gets to interact with the board. As opposed to the physical workshop, where participants can perform the actions at the same time, and the dialogue becomes a bouncing off of each other. By putting the PBO in a virtual context, we increase complexity in the collaboration which is something we wanted to reduce by making the boundary object in the first place.

Design suggestions for a VBO

- Reduce barriers between either facilitator or screen; i.e., by sticking to one virtual tool that serves both for communication and interaction.
- Enable interaction of participants across, between, or beyond the screen.
- Avoid ambiguity, as it may lead to a meta-conversation about the meaning and workings of the VBO.
- Limit the amount of choices; our PBO had three variables (pawn colour, weight, and positioning). A better option could be to limit to one choice or work with preset combinations.
- Enable a continuous tracking of movement of the BO by the participants to reflect changing states and relationships.
- Take care for the interface to be intuitive, so as to avoid discussion about how to use it. Here both an existing interface (emergent) or a designed interface could be workable depending on the familiarity of the participants with digital interfaces.

Our suggestions for dedicated design of a VBO echoes insights of antecedent literature in the workings of online collaboration. Virtual rooms need online collaboration tools designed to mitigate, e.g., the common problem of multitasking during virtual interactions (Karl et al., 2022) and to assist workshop participants in interpreting each other's non-verbal cues (Rae et al., 2015). While technology is continuously advancing, it is important to note that addressing these challenges requires more than just the platforms and tools themselves, but also asks for experienced online facilitators (Glikson et al., 2019).

Virtual sessions in Living Labs

On top of stressing a dedicated design for virtual collaboration spaces, we can conclude from our results that virtual communication, as it is now, inhibits boundary spanning capacity and a bridging of perspectives. In the virtual session, a boundary spanning did not occur on the topic of power behind AI. What did happen was that the participants tried to figure out the workings of the PBO through dialogue and evaluate its success. This could mean that a virtual session could be of value when testing a digital product, especially when participants are in different time zones or speak different languages. Therefore, we suggest that digital space nowadays is more conducive to other lab phases such as Challenge Definition and User Testing (related to Figure 1). In both these phases input from stakeholders with different perspectives are valuable. A virtual setting could increase the likelihood of participation of these stakeholders, provided that interpersonal connection and boundary spanning has already been established in the Dialogue phase. Of course, this hypothesis needs to be tested with further research.

We also already know from antecedent research that virtual teams encounter difficulties related to engagement, motivation, and trust (Choi & Cho, 2019; Karl et al., 2022; Morrison-Smith & Ruiz, 2020). Thus, building trust and interpersonal communication are crucial challenges for virtual collaboration to achieve performance, as these aspects influence the motivation and engagement of participants. It is suggested that the presence of experienced facilitators and technical support teams plays a critical role in making use of such technologies and, thus, ensuring a successful process. In our experiments we also noted an increased need of intervention by the online facilitator compared to the facilitator in the physical space. Moreover, Morrison-Smith & Ruiz (2020) highlight challenges in the sense of physical and psychological distance among participants. The fact that participants possess limited means to directly observe each other's contributions underscores the importance of facilitators aiding online group communication. This involves ensuring that quieter or more reserved individuals have a voice and moderating the involvement of more dominant participants.

Last but not least, we would like to add that choosing to work with a virtual or hybrid set-up is also a choice to share data of the conversation. Behind every virtual collaboration tool, such as Teams, Zoom, and Miro, there are companies who earn money from the data that is being shared. Hosts and facilitators need to be aware of data-sharing practices when considering to work with virtual collaboration tools.

Conclusion

The focus of this chapter was to find out how a virtual and physical facilitation of a physical boundary object (PBO) contributes to bridging perceptions and reframing multi-stakeholder dialogue in the Living Lab process. We compared the use of a PBO in the Dialogue phase of a Living Lab in physical space as opposed to virtual space. The work in this chapter is exploratory in nature. The results are not based on an extensive, systematic comparison of a number of workshops in both conditions. However, the work provides some interesting insights and points at promising directions for future research.

Most importantly, the results showed that the online experience was more fractured than the physical one. This makes it difficult for the multi-stakeholder participants to come to a common ground from which to develop an in-depth dialogue. The online dialogue was less about relationships between the different stakeholders and did not lead to formulating a common task. Observations of the physical experience, in contrast, showed that interacting with a boundary object in physical space supported bridging perceptions and reframing multi-stakeholder conversations to a more systemic level, resulting in finding a common task. This seems to indicate that direct feedback and physical interaction are important aspects of multi-stakeholder alignment in a complex systemic context. This resonates with work done about tangibility of objects in co-design research (Lockton et al., 2020). We can state that virtual communication inhibits boundary spanning capacity and a bridging of perspectives which is why it might not be advisable for the Dialogue phase of the Living Lab process. It might however be efficient for later phases such as the Challenge Definition or the User testing phases. To which extent our experiment within the AI context can be generalised to other contexts would need to be proven through further comparative research.

By comparing our results with antecedent literature on online (lab) collaboration, we also stress dedicated design to mitigate challenges and make six suggestions for the design of a virtual boundary object. Finally, we confirm that online collaboration requires experienced online facilitators and technical staff.

Future research

PBOs are a valuable tool to bridge different perceptions and determine a common task in a multi-stakeholder dialogue. While our analysis of two physical and virtual sessions underlines the importance of a physical meeting of people in a physical space in the Dialogue phase of a Living Lab for now, it would be interesting to experiment with virtual boundary objects in different lab contexts.

In future work, it would also be fruitful to look at combinations in the use of physical and virtual space for Living Labs (see also, Harbers & Overdiek, 2022). Departing from the Living Lab process from Figure 1, which phases need interaction in physical space? Why? And which phases could be effectively facilitated in virtual space and how? The suggestions provided in the discussion section could be a starting point for this future research.

Bibliography

- Alhojailan, M., & Ibrahim, M. (2012). Thematic analysis: A critical review of its process and evaluation. *International Journal of Qualitative Methods, 11*(1), 39–47.
- Carlile, P. R. (2004). Transferring, translating and transforming: An integrative framework for managing knowledge across boundaries. *Organization Science, 15*(5), 555–568. https://doi.org/10.1287/orsc.1040.0094
- Choi, O.-K., & Cho, E. (2019). The mechanism of trust affecting collaboration in virtual teams and the moderating roles of the culture of autonomy and task complexity. *Computers in Human Behavior, 91*, 305–315. https://doi.org/10.1016/j.chb.2018.09.032

- Cooney, R., Stewart, N., Ivanka, T., & Haslem, N. (2016). Design and the creation of representational artefacts for interactive social problem solving. In *Proceedings of the 2016 Design Research Society 50th Anniversary Conference*. Design Research Society.
- Dhawan, S. (2020). Online learning: A panacea in the time of COVID-19 crisis. *Journal of Educational Technology Systems, 49*(1). https://doi.org/10.1177/004723952093401
- Dietrich, T., Dalgaard Guldager, J., Lyk, P., Vallentin-Holbech, L., Rundle-Thiele, S., Majgaard, G., & Stock, C. (2021). Co-creating virtual reality interventions for alcohol prevention: Living lab vs. co-design. *Public Health Education, 9*. https://doi.org/10.3389/fpubh.2021.634102
- Glikson, E., Woolley, A., Gupta, P., & Kim, Y. J. (2019). Visualized automatic feedback in virtual teams. *Frontiers in Psychology, 10*. https://doi.org/10.3389/fpsyg.2019.00814
- Harbers, M., & Overdiek, A. (2022, January 7). Towards a Living Lab for responsible applied AI. In *Design Research Society*. DRS 2022, Bilbao.
- Haukipuro, L., & Väinämö, S. (2019). Digital user involvement in a multi-context Living Lab environment. *Technology Innovation Management Review, 9*(10). https://doi.org/10.22215/timreview/1273
- Karl, K. A., Peluchette, J. V., & Aghakhani, N. (2022). Virtual work meetings during the COVID-19 pandemic: The good, bad, and ugly. *Small Group Research, 53*(3), 343–365. https://doi.org/10.1177/10464964211015286
- Mačiulienė, M., & Skaržauskienė, A. (2020). Sustainable urban innovations: Digital co-creation in European Living Labs. *Kybernetes, 49*(7), 1969–1986.
- Miles, M. B., & Huberman, A. M. (1994). *Qualitative data analysis: An expanded sourcebook* (2nd ed.). Sage Publications.
- Morrison-Smith, S., & Ruiz, J. (2020). Challenges and barriers in virtual teams: A literature review. *SN Applied Sciences, 2*(6), 1096. https://doi.org/10.1007/s42452-020-2801-5
- Rae, I., Venolia, G., Tang, J. C., & Molnar, D. (2015). A framework for understanding and designing telepresence. In *Proceedings of the 18th ACM Conference on Computer Supported Cooperative Work & Social Computing*, 1552–1566. https://doi.org/10.1145/2675133.2675141

- Sartori, L., & Theodorou, A. (2022, January 24). A sociotechnical perspective for the future of AI: Narratives, inequalities, and human control. *Ethics and Information Technology, 24*(4).
- Star, S. L. (2010). This is not a boundary object: Reflections on the origin of a concept. *Science, Technology, & Human Values, 35*(5), 601–617. https://doi.org/10.1177/0162243910377624
- Star, S. L., & Griesemer, J. R. (1989). Institutional ecology, 'translations' and boundary objects: Amateurs and professionals in Berkeley's Museum of Vertebrate Zoology, 1907-39. *Social Studies of Science, 19*(3), 387–420.
- Thomas, E., Skeie, K. B., & Huang, H. (2022). Inclusion of elderly users via virtual spaces in the early stages of the innovation process. R&D Management, 54(2), 214-226. https://doi.org/10.1111/radm.12551
- Van der Horst, T., Overdiek, A., & Harbers, M. (2023). Designing a physical boundary object to invite dialogue about power relations behind AI systems. *Proceedings of Relating Systems Thinking and Design, RSD12.*
- Van Geenhuizen, M. (2018). A framework for the evaluation of Living Labs as boundary spanners in innovation. *Environment and Planning C: Politics and Space, 36*(7), 1280–1298. https://doi.org/10.1177/2399654417753623
- Van Veenstra, A. F., van Zoonen, E. (Liesbet), & Helberger, N. (2021). ELSA labs for human centric innovation in AI.
- Yasuoka, M., Akasaka, A., Kimura, A., & Ihara, M. (2018). Living Labs as a methodology for service design – An analysis based on cases and discussions from a systems approach viewpoint. *International Design Conference – Design*, 127–136. https://doi.org/10.21278/idc.2018.0350

Tomasz Jaśkiewicz, Iskander Smit

10. Between Experiments – Leveraging Prototypes to Trigger, Articulate, and Share Informal Knowledge

Case of the Cities of Things Living Lab

DOI: 10.1201/9781003491484

Introduction

Living Labs are 'open innovation ecosystems' (ENoLL, 2024), offering unique opportunities for conducting scientific experiments 'in the wild' (Romero Herrera, 2017; Ballon & Schuurman, 2015). Yet, Living Labs also support the generation of other, informal, types of knowledge: often tacit and difficult to capture and share, acquired outside of rigorous academic research frames (Schuurman & Tõnurist, 2017), and comprising insights, ideas and know-how gained through experience, serendipity, and sometimes ad-hoc and playful explorations (Jaskiewicz, 2021). This chapter focuses on the significant, yet often overlooked, role that prototypes play in triggering, articulating and sharing such informal knowledge. Grounded in the discourse on the role of prototypes in Research through Design (RtD), the chapter brings forth that prototypes are not just valuable as artefacts to be studied through formalised research but are also crucial in supporting the generation of rich, contextual insights, ideas and know-how, sharing them across experiments, disciplines and stakeholders, while often facing the challenge of legitimisation and generalisation of such knowledge.

The Cities of Things Living Lab in Rotterdam serves as a case study to investigate and illustrate the roles prototypes play in triggering, articulating, and sharing informal knowledge. The lab's focus is on investigating the opportunities and threats of emergent digital technologies, especially urban robotics, on everyday life in urban neighbourhoods. Over the 24-month period of lab operation, which we have coordinated, various, often unstructured, and sometimes spontaneous and playful, co-creation sessions, workshops and hackathons gave rise to over 30 'civic robot' (Jaskiewicz, 2022) prototypes and provided us with insights into how these artefacts can facilitate the integration of diverse viewpoints, expertise, and knowledge from local communities and stakeholders. These prototypes have proven themselves to be instrumental in generating new ideas, perspectives, critical reflections and even potential business models, in this way highlighting the value of iterative prototype creation as a mechanism for knowledge sharing in the context of Living Labs.

Research Through Design, with Prototypes, in Living Labs

Living Labs, RtD and prototyping are related concepts. Living Labs are user-focused experimental environments where users, experts and stakeholders co-create innovative solutions in real-life settings (e.g., Steen & Van Bueren, 2017). RtD is a research approach that leverages design practice as a way of generating and communicating new knowledge (e.g., Stappers & Giaccardi, 2017), which can be used in the Living Lab setting (Keyson et al., 2017). Prototyping is a process of creating and testing tangible representations of new ideas, products, or services (e.g., Camburn et al., 2017), and prototypes typically play a central role as objects of enquiry in both Living Lab and RtD processes.

Engaging users and other stakeholders in co-creating innovative solutions in real-life settings is central to Living Lab approaches (Følstad, 2008; Leminen et al., 2012; Ballon & Schuurman, 2015; Steen & Van Bueren, 2017). The discourse on RtD offers a broader perspective on the interplay of design and knowledge generation processes enabled by co-creation, and provides some insights and guidelines for conducting and communicating design-based research in different contexts and domains (Frayling, 1993; Zimmerman et al., 2007; Koskinen et al., 2013; Stappers & Giaccardi, 2017). RtD is an approach originating from design practice, and substantial discourse on RtD is dedicated to mandating it as academic or scholarly activity (Cross, 2001; Koskinnen et al., 2013; Sanders & Stappers, 2014), whereas it may also serve as a way to provide additional insights enhancing academic research (Höök & Löwgren, 2012). In this context, the iterative character of RtD is a key distinguishing factor from many established research practices, where RtD involves moving back and forth between design and research-oriented activities, sometimes referred to as problem and solution spaces (Dorst & Cross, 2001), or concept and knowledge spaces (Hatchuel & Weil, 2009), heavily influenced by Donald Schön's seminal work on reflection in- and on-action in design processes (Schön, 1983), as shown in Figure 1.

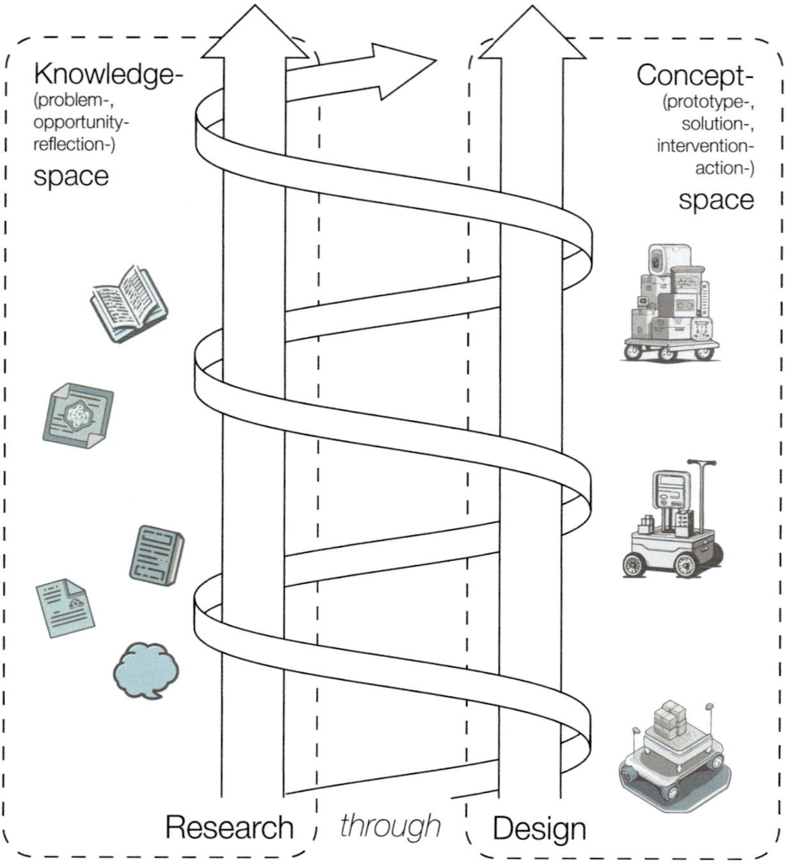

Figure 1. Research through Design process can be represented as an iterative trajectory between knowledge and concept spaces, where prototypes enable generation of new knowledge, which in turn leads to creation of new prototypes.

Prototypes are often central to both Living Lab and RtD processes. While typically defined as tangible representations of new ideas made to be evaluated and tested, they are also recognised in their role as boundary objects (Star, 1989) supporting articulation and exchange of knowledge in multidisciplinary teams and communities (Lim et al., 2008). In Living Lab settings, but also in most RtD practices, prototypes become embedded in real-world environments, which extends their outreach. It is through prototypes, their histories, experiences and anecdotes that most Living Labs and many researching designers engage with their context of inquiry and communicate their research findings to the outside world (Pallot et al., 2010). The literature on prototyping provides useful insights and guidelines for creating and using prototypes in different design stages and research activities (e.g., Lim et al., 2008; Snyder, 2003). While early literature focuses on prototyping in controlled or laboratory settings, where prototypes are mainly used for testing and validating design concepts or hypotheses (e.g., Houde & Hill, 1997), there has been a growing focus on prototyping in real-world or naturalistic settings, where prototypes are embedded in the actual use context and exposed to the complexity and dynamics of the real world (Buchenau & Fulton-Suri, 2000). The discourse on how prototypes support triggering, articulating and sharing informal knowledge that emerges from the real-world interactions and experiences of users and other stakeholders is long-standing (e.g., Gaver, 2012), but it remains broad and open-ended.

Prototypes in the Cities of Things Lab

The Cities of Things Lab is a Living Lab that focuses on smart city innovation, with a particular focus on investigating the societal impact and opportunities of introducing 'smart city things', especially 'city bots'[1] (which include, for example, autonomous delivery robots, self-driving street-cleaning machines, or mobile security cameras) in the context of the city of Rotterdam, in relation to various urban themes, such as mobility, safety, health, circularity, or climate action. The lab has been established as a collaboration between the Cities of Things Foundation, the Research Centre Creating 010 at the Rotterdam University of Applied Sciences, Afrikaanderwijk Cooperative, Studio for the City network of design agencies, ThingsCon, and the Innovation Network 'VONK' of the City of Rotterdam. Besides these partners, it involves a broader group of stakeholders, such as citizens, researchers, companies, and public organisations, in co-creating and testing innovative 'smart city things'. The lab has multiple objectives. First, it aims to make citizens aware of the possible futures in which they will be living, working and recreating in interplay with 'smart city things' and to empower citizens to take an active role in shaping those futures, particularly in the context of their own neighbourhood.

1 We use the term 'smart city things' to refer to a broad category of artefacts that are tangible touchpoints with the digital infrastructure of the so-called 'smart cities'. Within that category, 'city bots' are robotic operating devices with an urban-related function. The later introduced 'WijkBots' are city bots operating in neighbourhood contexts, relating to the local daily life of citizens.

Second, the lab aims to contribute to establishing Rotterdam as a frontrunner and innovative city of coexistence of 'smart city things' and people. Third, it aspires to strengthen design businesses in Rotterdam in their 'smart city thing' design knowledge and skills and thus strengthen their competitiveness and innovation capacity. Fourth, it aims to develop and popularise methodologies for engaging citizens in contributing the local expertise and knowledge in shaping the 'smart city'. Finally, the lab also shares with citizens and other stakeholders (for example, neighbourhood organisations or the municipality of Rotterdam) the knowledge and experience relating to urban robotics in order to perpetuate the achieved innovations and support them through policy and regulation.

As part of work in the Lab, we have used the WijkBotKit platform (where 'wijk' means 'neighbourhood' in Dutch; see Figure 2; WijkBot.nl, 2024) to create and use prototypes. The platform is a collection of open-source, low-cost tools and methods for co-creating 'city bots' that can be designed, developed, owned, maintained and operated by local communities. The kit is based on hacked 'hoverboards' (self-balancing, two-wheeled vehicles popular with kids) and off-the-shelf components. It allows non-experts to jointly build full-scale, remote-controlled robot prototypes in a few hours. A participant can develop and then operate the prototype as a 'wizard-of-Oz' (Dahlbäck et al., 1993), controlling the robot's movements, optionally talking through its speakers, listening and watching through its microphones and cameras. The WijkBotKit is a platform in permanent development. With the help of students at the Rotterdam University of Applied Sciences, the kit has also been extended with modules enabling increased autonomy of the prototyped robots. This includes, for example, modules for autonomous navigation, space mapping, obstacle avoidance, or services for managing multiple robots operating in a neighbourhood, and it enables the development of prototypes that are closer to actual end-products, and that can be put to use for longer test periods with less involvement of a humanoperator. Instructions and documentation of the WijkBotKit have been released as open-source materials at: www.wijkbot.nl.

Figure 2. The WijkBotKit platform in a workshop involving co-creation of a 'city bot' (left), and its prototype tryouts (right).

Co-creation Activities

We have used the WijkBotKit platform in three different types of co-creation activities within the Living Lab: (1) a series of nine 'think-tank' co-creation sessions with a persistent group of citizens of Rotterdam, (2) workshops and hackathons with one-off participants, typically as part of other events such as festivals or conferences, and (3) student assignments in the Communication and Multimedia Design, Technical Informatics and Industrial Product Design programmes of the Rotterdam University of Applied Sciences, and master's programme Next Level Engineering of The Hague University of Applied Sciences. In each of these activities, we have engaged the participants in co-creating and trying out prototypes of city bots. The co-design sessions, workshops, and hackathons were facilitated by researchers and designers from the Living Lab team, while the student assignments were supervised by the University of Applied Science teachers. The duration, frequency, and location of the activities varied depending on the type and the goal of the activity.

Specifically, the 'think-tank' sessions took place in the Afrikaanderwijk, a multicultural neighbourhood in Rotterdam, and were attended by a group of nine citizens. The shared aim of the sessions was to explore and prototype city bots that could address the urban challenges and opportunities in the neighbourhood identified by the participating citizens, for example, waste management, social cohesion, and cultural diversity. The sessions were organised at the community house of Afrikaanderwijk Cooperative. A typical session would start with a demonstration of the earlier and reference work and findings. In the case of the first session, it was a demonstration of the existing city bot examples and the first version of the WikBotKit. The sessions would then follow different workshop formats and use a selection of context-mapping tools (Visser et al., 2005) to articulate participants' views on the neighbourhood and identify opportunities and threats relating to introducing city bots in relation to these views, and formulating prototype ideas. Subsequently, the WijkBotKit was used to create prototypes of these ideas, imitating prototyping during sessions, and continuing it with available participants in the following days.

To collect and analyse data from the 'co-design think-tank' sessions, we have used a mixed-methods approach, with a strong focus on qualitative data. The data sources included: (a) protocols of sessions, which documented the main topics, decisions, and outcomes of each activity in the form of proceedings or annotated timelines, (b) photos and video recordings, which captured the results of the co-creation and other activities, (c) document analysis, which examined the documentation and prototypes produced by the participants, and (d) direct observation, which involved researchers and designers from the Living Lab team participating and observing the activities. The data analysis differed per activity and followed an inductive and iterative process, using among others thematic coding and content analysis to identify and categorise the main themes, patterns, and insights emerging from the collected data. Once operational, the prototypes would be tried out in their intended context of use, observed and documented through photographs, video recordings and journaled observations.

In parallel to the above, more structured activities, the WijkBotKit was applied in other situations, such as workshops and student projects. These activities resulted in other, various prototypes of city bots. Some examples of those are:

- A one-day workshop at City Climate meets Creative Coding festival in Hamburg, where the participants co-created and tested robot prototypes for dealing with climate change consequences in the city.
- A student assignment, where the students co-created and tested prototypes supporting urban gardening.

These show the versatility and the potential of WijkBotKit in co-creating and testing city bots in different domains and contexts. They delivered valuable input and prototypes in the process, but were not studied rigorously due to organisational constraints.

Evolution of Prototypes

In the course of the Living Lab's operation, more than 30 different prototypes were built with varying fidelity, purpose and context of use (Figure 3). The prototypes influenced one another, as each co-creation activity would normally start with a demonstration of prior prototypes, and would end or be followed by a co-prototyping activity. During each session, participants would explore different ideas for new applications of city bot technology, typically using sketches, post-its, Lego blocks and other creative materials (Figure 4). After clustering types of applications, the city bot functions would be decided in a joint group discussion, taking into account what contributes most to the challenge or context at hand. Consequently, the prototype would be built using WijkBotKit and recycled industrial waste scrap materials.

Figure 3. The WijkBotKit has been used to create more than 30 prototypes to date.

Figure 4. Cities of Things Lab involved co-creation sessions with citizens using mixed materials to express ideas and make decisions.

The WijkBotKit prototypes evolved and developed different strands over the 24 months of the different Living Lab co-creation activities. Figure 5 illustrates the chronological timeline of prototypes, project activities and thematic focus, such as circularity, social features or transport that emerged from the co-prototyping process. The WijkBotKit platform enabled the participants in different types of sessions to experiment with different forms and functions of the city bots, and to explore different scenarios and contexts of use, from neighbourhood-focused scenarios to personal services. In turn, the kit evolved alongside the project with features being influenced by project activities. The participants were able to express their personal preferences, values, and emotions through the prototypes, and to interact with them in both playful and meaningful ways. The different prototypes showed a range of scenarios and functions, such as community garden coach, a street ambulance for homeless people, a waiter robot, or a robotic climate advocate. Through these kinds of activities, the prototypes appeared to unlock an embodied dynamic and collaborative learning process and served as boundary objects, as they opened up multiple interpretations of what smart city things can mean, link different social worlds and while being context-specific, facilitating communication within and outside of the co-creation process with participants and outside stakeholders. The prototypes also had ripple effects between sessions, enabling iterating between understanding and articulation of needs and new ideas, perspectives, and even business ideas for applying the WijkBotKit in new contexts.

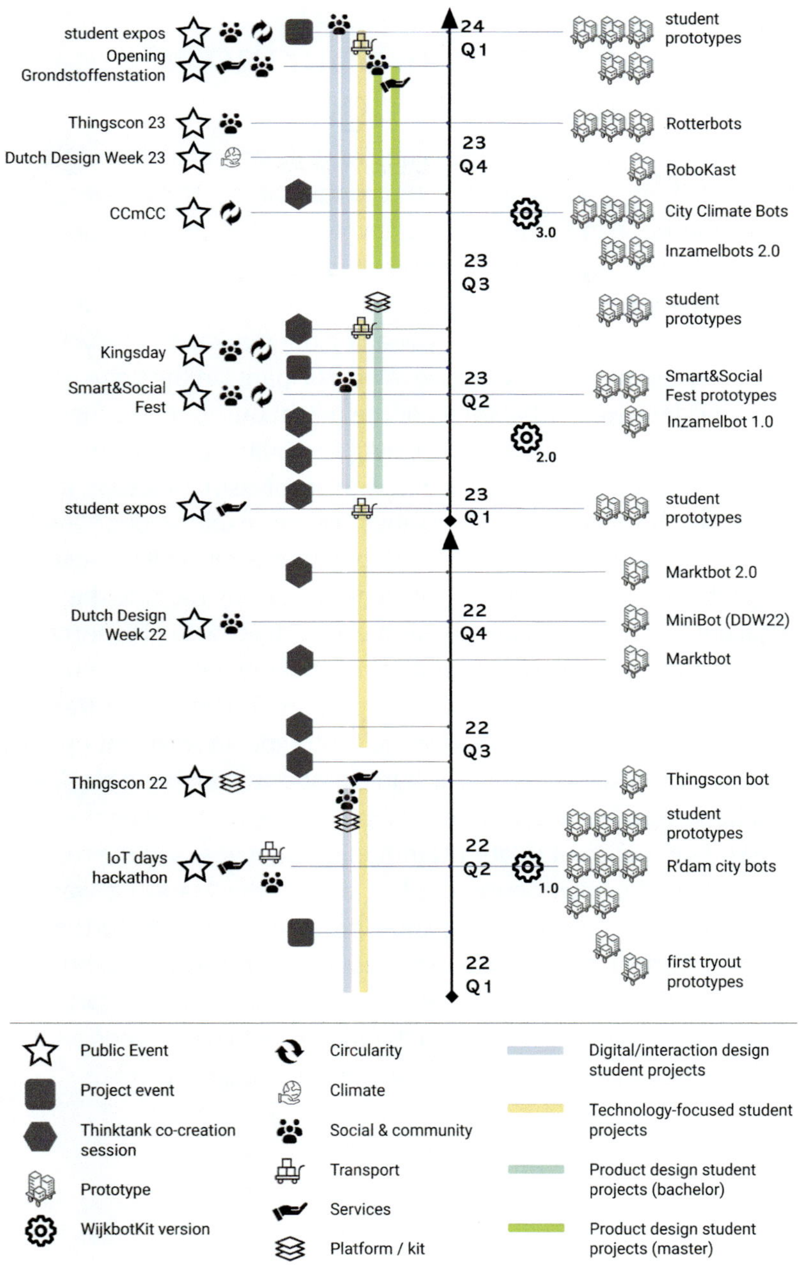

Figure 5. Chronological timeline of prototypes, project activities and thematic focus in the Cities of Things Lab summarises and shows the intricacy involved in tracking sources of informal knowledge emerging across the lab.

Example of Triggering, Articulating, and Sharing Informal Knowledge

The 'think-tank' co-creation sessions were a particular case of the Cities of Things Lab focused on the context of the Afrikaanderwijk neighbourhood in Rotterdam and involving a community of its inhabitants. Afrikaanderwijk is known for its large market, and the community-run Afrikaanderwijk Cooperative is involved in multiple initiatives dealing with cleaning the market and recycling waste, including unsold produce. All participants in the 'think-tank' sessions were inhabitants of the neighbourhood. During the first session, the idea of designing and building robots for the neighbourhood was generally met with scepticism. However, it eventually became subject to various creative explorations. One of the first ideas was to create a 'Marktbot' that would collect unsold vegetables and fruits from the local market, bring them to the adjacent 'Grondstoffenstation'[2] and use such collected resources in the community kitchen operating in a neighbouring community house, where each week volunteers prepare free meals for those in need. Discussing the ideas for the prototype has brought up the different civic values that participants associated with the market, neighbourhood, and community life. It also helped explain the workings of different related community initiatives, and the roles they play in the neighbourhood. The idea for the 'Marktbot' was a manifestation of these values, and a practical attempt to further support these initiatives. After the session, a prototype was built by mounting the crates used at the market on the WijkBotKit cart, and it was subsequently tried out on the market, while equipped with a 360-degree camera recording anonymised footage (Figure 6).

2 The Grondstoffenstation (resource station) was officially jointly opened by the council member of Rotterdam and one of the WijkBots on 10 January 2024 (Linkedin post by City of Rotterdam).

Figure 6. The Marktbot triggered many unexpected situations and discussions during a try-out at the local market in the Afrikaanderwijk neighbourhood of Rotterdam.

During its trial run at the market, the robot subtly merged into the scene, initially not drawing much attention. Within minutes, one of the market vendors intuitively put a crate with stale bread on the robot, hinting that the robot's purpose was being understood despite its discreet presence. Analysis of the robot's camera footage later revealed its true impact: it sparked curiosity and interaction, gently pulling people together to wonder about its purpose. A particular instance saw a father and daughter following the Marktbot for half-an-hour, mingling and musing with others about the enigmatic bot. The robot wove itself into the social fabric, fostering social interactions and discussions. As the day progressed, reactions to the robot shifted from delight to annoyance, especially from a vendor concerned about privacy and his own dubious activities. These varied responses highlight the complex roles civic robots play in urban environments, evoking a wide array of human reactions based on the situation.

The next 'think-tank' co-creation session in the Afrikaanderwijk started with a shared reflection on the above experiences. The participants were inspired to initiate a new endeavour that would enhance the involvement of robots in urban teamwork. They conceived an idea of a robot that could gather not just physical materials like paper, plastic, and metal but also collect the intangible: stories, thoughts, and reactions from people around. This idea stemmed from considering the prior prototype's influence, driving the group to imagine a more 'social' robot that could interact more with residents, sparking dialogue, and encapsulating the communal spirit of the area. It led to shifting participants' perspective from seeing a city bot as a purely functional device to recognising its role in community interactions, fostering human contact, and showing its potential as catalyst of new forms of community interactions. Earlier, we have extended the dimensions of the WijkBotKit so that it could support building bigger robots, and use pneumatic wheels for better outdoor operation. This together led to the next prototype, which was named 'Inzamelbot' (collecting bot). The Inzamelbot has been put to use during a neighbourhood festival[3] (Figure 7) and has since had multiple other versions and follow-ups. The Marktbot and Inzamelbot focused on neighbourhood interactions around the market. For example, engaging with the role and relationship between humans and robots in the smart city context. In the Inzamelbot case, the notion of the relation of the city bot would be shifted from an autonomous operating tool to a buddy for human workers, their shared agency in performing both cleaning and social activities in the neighbourhood.

The latter aspect has been a returning discussion topic after experiencing the interaction with a prototype, including who controls and helps whom, and how to design for co-performance and collaboration between both types of actors, exemplifying the ongoing academic discourse on the topic (Kuijer & Giaccardi, 2018). Different lab prototypes triggered discussion on various facets of this challenge. For example, one of the first city bots developed during the Rotterdam IoT day hackathon was a de-escalating bot for moments of tensions in groups on the street, where a robot would proactively enter into a difficult social situation as a third party. Another example is the student-made community garden bot 'Moesie', which raised the question if the smart city thing needs to take the lead or even take over tasks of a social intermediary in urban communities. Yet another student group developed a city bot that could assist elderly people with their daily activities, such as shopping, gardening, or navigating through the city. The students considered how to make the WijkBot responsive and adaptive to the needs and preferences of the elderly users, and how to create a sense of trust and companionship between them, by defining how the robot would lead the way in planning an inspiring shopping experience, follow or guide the elderly in its role as a motorised walking and shopping aid. The students brought

3 The festival was organised as part of the celebration around the visit of the Dutch king to the neighbourhood on 27 April 2023.

the prototype to the streets of the city and discussed with elderly passersby the potential challenges and risks of relying on a robot for assistance, and how to ensure the safety and autonomy of elderly citizens through such human-robot partnership.

Figure 7. The 'Inzamelbot' prototype was built by citizens of the Afrikaanderwijk using the WijkBotKit platform (left), and put to use to 'collect garbage and stories' (right), 'evolving' some ideas from the earlier Marktbot prototype in Figure 6, but becoming in fact a new city bot 'species'.

Learning with Prototypes

The co-prototyping activities in Afrikaanderwijk exemplify how the co-creation and trying out of resulting prototypes provides a fertile ground for the convergence and divergence of knowledge, skills, and perspectives. Designing and building prototypes converged knowledge by combining insights from our earlier research with experiences and abilities of the participants. The tryout of the robot on the market diverged the knowledge by revealing the richness of social engagement with robots in the city, sparking many unexpected discussions, and triggering many new insights and ideas. The Cities of Things Lab has enabled tens of such iterations, where citizens would often bring in new knowledge, ideas, and know-how, and, through a facilitated co-design process, would translate this knowledge into the prototype design. In turn, the prototypes perpetually enabled the confrontation of this synthesised knowledge with the world, leading to new ideas, insights and know-how. In each of these iterations, from the perspective of informal knowledge generation, we have identified three distinct steps of engaging with the prototypes, listed below and illustrated in Figure 8.

1. Triggering Knowledge

The encounters with prototypes during prototype try-outs served as a spark for co-creation participants and passers-by alike to reflect on the future of their neighbourhood and their personal perspectives towards it. By interacting with the prototype while being embedded in the actual context of the neighbourhood, people were triggered to relate the possible futures embodied by prototypes to their everyday-life situations, rather than think in abstract terms.

2. Articulating Knowledge

As the prototype facilitated spontaneous discussions, it also became a boundary object that supported the articulation of people's often informal, tacit knowledge in relation to its function, role, form and potential in the context of people's everyday lives. Leveraging the prototype as a common reference, people oftentimes engaged in deep conversations about the community's values, challenges, and aspirations, effectively using the prototype to articulate their personal understanding of the neighbourhood challenges and opportunities. By asking questions about the prototypes, trying to explain what they were or by criticising them, they were prompted to explain their personal, often abstract, reflections on the roles of technological innovations in the city through concrete examples and situations.

3. Sharing Knowledge

Participants and passers-by would end up sharing and discussing their insights and experiences rooted in the community's practices. After these conversations took place, the prototypes would be reminders of the discussed topics, serving as easy to explain, intriguing examples to bring up in later communications. These communications included presentations, exhibitions or publications by lab members and affiliates, but also became featured in informal conversation, social media posts or popular media made by people oftentimes not directly involved in lab's activities. Especially in the latter, the prototype itself would be the main topic or conversation-starter, and associated insights, ideas and know-how would follow. In these ways, the prototypes not only facilitated the articulation and transfer of tacit knowledge but also further exchange and development of insights, ideas and know-how by engaging broader audiences, that the following co-creation sessions would gather and take as input to build upon further.

Notably, the prototypes would regularly serve as cases supporting explanation of knowledge obtained through lab research in academic contexts (such as scientific articles or presentations), while being the leading topic of lab-related communication in the popular context (such as social media posts, popular media articles, or told stories).

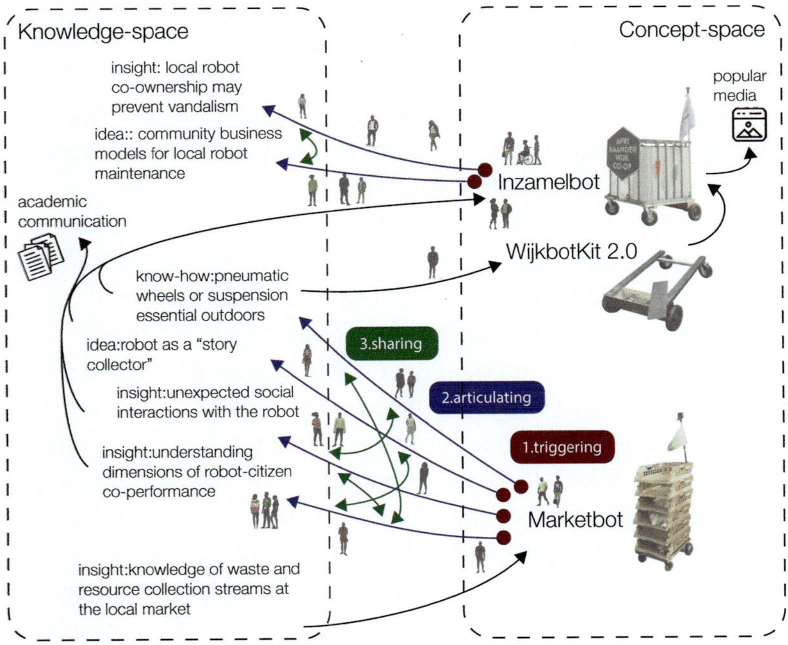

Figure 8. Transition from Knowledge- to Concept-space is often a complex process, where prototypes trigger (often unexpected) people to articulate their insights, ideas and know-how and share with each other, ultimately leading to a new prototype.

The initiatives that followed were not restricted to what was organised or curated by the Living Lab, and they included several spin-offs. One example is the ongoing development of an educational game by one of the citizen participants. The participant had prior affinity with STEM school education and contacts with local NGOs operating in this field. The game he conjured up and aspires to further develop is intended to empower children and their parents to envision and actively shape desirable futures with urban robots, while promoting STEM interests among the youth.

Discussion

Across the discussed examples from the Cities of Things Lab, prototypes come across not just as tools for practical problem-solving but as vital instruments for supporting shared understanding and collaborative community efforts, highlighting the plurality of citizens' roles, perspectives, and interpretations in the manifestation, clarification, and dissemination of informal knowledge within community settings. In the examples, we focused on the informal aspects of knowledge generation that happened in parallel to more rigorous research activities. Our investigation shows how the city bot prototypes brought together diverse expertise, facilitating collaborative knowledge and skill convergence. While participants contributed their unique knowledge perspectives and know-how, the prototypes became tangible manifestations of their collective intelligence. The prototypes also made it possible to explore different aspects of smart city things, both on functional and social interactions between humans and robots. This journey of co-creation of city bots unveils the transformative power of prototypes in fostering serendipitous learning, social encounters, and collaborative creation. This process, deeply rooted in local contexts, offered broad, and often unexpected insights into the potential of prototypes as boundary objects that catalyse social innovation and envision the future of urban living. Through collaborative endeavours, such as the development of an educational game by a citizen participant, the lab's explorations show the significance of future-thinking and technological literacy within local communities. This supports developing a shared vision and inclusive dialogue around urban robotics, highlighting prototypes not just as research artefacts but as crucial tools in generating, articulating, and disseminating rich contextual insights and know-how.

The co-prototyping activities in Afrikaanderwijk exemplify how engaging with prototypes catalyses the convergence and divergence of knowledge, skills, and perspectives. Prototypes served here as tangible manifestations of collective intelligence, enabling participants to explore different aspects of smart city solutions and foster new forms of community interaction. Clearly, such process also has substantial imitations, which need to be acknowledged and addressed. The knowledge and insights from our examples were scattered, subjective, and superficial, as they depended on the specific situations, participants' personal opinions, and unpredictable interactions. The process was by its nature unpredictable, messy and difficult to capture, as it involved multiple steps, activities, prototypes and experiments, with varying degrees of documentation and evaluation being attainable or at all possible. The resulting insights are largely anecdotal, subjective, difficult to compare, validate and generalise, and very time-consuming to gather and document.

While this chapter focuses on informal knowledge tiggered, articulated and shared with help of prototypes, it is not our intention to diminish the value and role of rigorous, structured research activities in Living Labs. To the contrary, we aspire to open up a methodological discourse on how these two could better complement each other, including exploring the middle ground between them. To this end, we continue to investigate various tools and methods for documenting and evaluating prototypes and their iterations. This includes the use of annotated portfolios, structured reflection cards, and designer logbooks to capture and articulate the insights generated, suggesting new directions for research and innovation in smart city domains (Jaskiewicz, 2022).

As we reflect on the lessons learned and the paths forward, it is clear that the future of technological innovations such as urban robotics demands a collaborative, inclusive, and reflective approach. Engaging in policy-level discussions, addressing ethical considerations, and exploring governance-related, legal, and societal themes are some of many essential steps in ensuring that the integration of technology into our everyday lives has the potential to enhance, rather than diminish, our social experience. The examples from the Cities of Things Lab not only showcase the potential of civic robotics as an example of such technology to transform urban spaces but also invite us to continue the conversation, experimentation, and action towards creating more connected, resilient, and inclusive cities. Prototypes are here not just valuable as artefacts to be studied through formalised research but are also crucial in supporting the generation of rich, contextual insights, ideas and know-how, sharing them across experiments, disciplines and stakeholders while often facing the challenge of legitimisation and generalisation of such knowledge.

Conclusion

In this chapter we explored how prototypes help generate, articulate, and share informal knowledge in Living Labs, using the Cities of Things Lab as a case study. We discussed the co-prototyping process and the role of prototypes as convergence catalysts, and we illustrated the discussion with examples from the Cities of Things Lab. We also reflected on the challenges and lessons learned from the co-prototyping process, and we proposed a framing that conceptualises prototypes as knowledge convergence catalysts in iterative research through design (RtD) processes.

Informal ways of learning and sharing knowledge through prototypes are relevant for the Living Lab methodology, and in this chapter, we also emphasised the importance of RtD as a theoretical frame for capturing such knowledge. We advocate for the development of new research methods that fit in this frame to further enable longitudinal studies and comparisons of innovation processes, while leveraging the role of prototypes and other boundary objects to trigger, articulate and share informal knowledge. Finally, we also touched upon the value of playfulness, open-endedness and serendipity in this approach, bringing people together, and engaging them in creative ways to not only generate new knowledge, but also to act upon this knowledge for common good.

Bibliography

- Almirall, E., Lee, M., & Wareham, J. (2012). Mapping Living Labs in the landscape of innovation methodologies. *Technology Innovation Management Review, 2*(9).
- Ballon, P., & Schuurman, D. (2015). Living Labs: Concepts, tools and cases. *info, 17*(4).
- Bouwma, I., Wigboldus, S., Potters, J., Selnes, T., van Rooij, S., & Westerink, J. (2022). Sustainability transitions and the contribution of Living Labs: A framework to assess collective capabilities and contextual performance. *Sustainability, 14*(23), 15628.
- Brown, T. (2009). *Change by design: How design thinking transforms organizations and inspires innovation*. HarperCollins.

- Buchenau, M., & Fulton-Suri, J. (2000). Experience prototyping. In *Proceedings of the 3rd conference on Designing interactive systems: Processes, practices, methods, and techniques* (pp. 424-433).
- Camburn, B., Viswanathan, V., Linsey, J., Anderson, D., Jensen, D., Crawford, R., ... & Wood, K. (2017). Design prototyping methods: State of the art in strategies, techniques, and guidelines. *Design Science, 3*, e13.
- Cross, N. (2001). Design/science/research: Developing a discipline. In *Proceedings of the Korea Society of Design Studies Conference* (pp. 16-24). Korea Society of Design Science.
- Dahlbäck, N., Jönsson, A., & Ahrenberg, L. (1993, February). Wizard of Oz studies: Why and how. In *Proceedings of the 1st international conference on Intelligent user interfaces* (pp. 193-200).
- Dorst, K. (2015). *Frame innovation: Create new thinking by design*. The MIT Press.
- Dorst, K., & Cross, N. (2001). *Creativity in the design process: co-evolution of problem–solution*. Design studies, 22(5), 425-437.
- European Network of Living Labs. (2024, February 16). *Home – European Network of Living Labs, Living Labs network*. https://enoll.org/
- Følstad, A. (2008). Living Labs for innovation and development of information and communication technology: A literature review.
- Frayling, C. (1993). Research in art and design. *Royal College of Art Research Papers, 1*, 1-5.
- Gaver, W. (2012). What should we expect from research through design? In *Proceedings of the SIGCHI conference on Human factors in computing systems* (pp. 937-946).
- Hatchuel, A., & Weil, B. (2009). CK design theory: An advanced formulation. *Research in Engineering Design, 19*, 181-192.
- Höök, K., & Löwgren, J. (2012). Strong concepts: Intermediate-level knowledge in interaction design research. *ACM Transactions on Computer-Human Interaction (TOCHI), 19*(3), 1-18.
- Jaskiewicz, T., & Van Der Helm, A. (2018). Unlocking the interactive office: Concurrent prototyping approach. In *Proceedings of the 2018 Designing Interactive Systems Conference* (pp. 547-558).

- Jaskiewicz, T., Mulder, I., Verburg, S., & Verhij, B. (2018). Leveraging prototypes to support self-directed social learning in makerspaces. In *DS 93: Proceedings of the 20th International Conference on Engineering and Product Design Education (E&PDE 2018)*, Dyson School of Engineering, Imperial College, London, 6th-7th September 2018 (pp. 430-435).
- Jaskiewicz, T. J. (2022). Civic prototyping: A creative encounter between design prototypes and engaged citizens. *Inaugural Lecture, Hogeschool Rotterdam*.
- Jaskiewicz, T. (2022). Learning from prototypes: From the design studio to the city. In *Applied Design Research* (pp. 43-52). CRC Press.
- Keyson, D. V., Morrison, G. M., Baedeker, C., & Liedtke, C. (2017). Living Labs to accelerate innovation. In *Living Labs: Design and Assessment of Sustainable Living* (pp. 55-61).
- Koskinen, I., Zimmerman, J., Binder, T., Redstrom, J., & Wensveen, S. (2013). Design research through practice: From the lab, field, and showroom. *IEEE Transactions on Professional Communication, 56*(3), 262-263.
- Kuijer, L., & Giaccardi, E. (2018, April). Co-performance: Conceptualizing the role of artificial agency in the design of everyday life. In *Proceedings of the 2018 CHI Conference on Human Factors in Computing Systems* (pp. 1-13).
- Leminen, S., Westerlund, M., & Nyström, A. G. (2012). Living Labs as open-innovation networks.
- Lim, Y. K., Stolterman, E., & Tenenberg, J. (2008). The anatomy of prototypes: Prototypes as filters, prototypes as manifestations of design ideas. *ACM Transactions on Computer-Human Interaction (TOCHI), 15*(2), 1-27.
- Lupetti, M. L., Smit, I., & Cila, N. (2018). Near future cities of things: Addressing dilemmas through design fiction. In *Proceedings of the 10th Nordic Conference on Human-Computer Interaction* (pp. 787-800).
- Martin, B., & Hanington, B. (2012). *Universal methods of design: 100 ways to research complex problems, develop innovative ideas, and design effective solutions*. Rockport Publishers.
- Pallot, M., Trousse, B., Senach, B., & Scapin, D. (2010, August). Living lab research landscape: From user centred design and user experience towards user cocreation. In *First European Summer School 'Living Labs'*.

- Romero Herrera, N. (2017). The emergence of Living Lab methods. In *Living Labs: Design and Assessment of Sustainable Living* (pp. 9-22).
- Sanders, E. B. N., & Stappers, P. J. (2014). Probes, toolkits and prototypes: Three approaches to making in codesigning. *CoDesign, 10*(1), 5-14.
- Schön, D. A. (1983). *The reflective practitioner: How professionals think in action*. Basic Books.
- Schuurman, D., & Tõnurist, P. (2017). Innovation in the public sector: Exploring the characteristics and potential of Living Labs and innovation labs. In *OpenLivingLab Days 2016* (pp. 78-90).
- Snyder, C. (2003). *Paper prototyping: The fast and easy way to design and refine user interfaces*. Morgan Kaufmann.
- Stappers, P. J., & Giaccardi, E. (2017). Research through design. In *The Encyclopedia of Human-Computer Interaction* (pp. 1-94). The Interaction Design Foundation.
- Star, S. L., & Griesemer, J. R. (1989). Institutional ecology, translations, and boundary objects: Amateurs and professionals in Berkeley's Museum of Vertebrate Zoology, 1907-39. *Social Studies of Science, 19*(3), 387-420.
- Steen, K., & van Bueren, E. (2017). The defining characteristics of urban Living Labs. *Technology Innovation Management Review, 7*(7).
- Visser, F. S., Stappers, P. J., Van der Lugt, R., & Sanders, E. B. (2005). Contextmapping: Experiences from practice. *CoDesign, 1*(2), 119-149.
- Cities of Things Foundation. (2024, February 16). *Cities of Things Lab 010 WijkBotKit*. https://www.Wijkbot.nl
- Zimmerman, J., Forlizzi, J., & Evenson, S. (2007, April). Research through design as a method for interaction design research in HCI. In *Proceedings of the SIGCHI Conference on Human Factors in Computing Systems* (pp. 493-502).

Guido Stompff, Mark Jacobs,
Donagh Horgan

11. Ceci n'est pas un Prototype

Prototypes and Idealtypes as Representations of What Works and What Matters

DOI: 10.1201/9781003491484

Summary

Prototypes are omnipresent in design (thinking) practice and related literature. In this chapter what a 'prototype' represents and what it contributes to design processes is explored. A distinctive lens on experimental environments is adopted, where individuals without design expertise engage in co-design. The analytical framework positions design activities as three interrelated inquiries, inquiries into the existing situation (what is?), into the ideal situation (what is preferred?), and into plans of action (what can be?). The study, set in several Urban Living Labs in which students engage, revealed consistent patterns: novices to design thinking have a preoccupation with the ideal, are reluctant to express what is on their mind, and as a result hardly iterate.

The findings suggest that what a prototype represents and which type of inquiry it serves differs among stakeholders, often resulting in unmet expectations. For some, it should advance the inquiry into what is preferred, that is: what matters? For others, it should advance the inquiry into what can be, that is, what realistically can be achieved? Both inquiries are valuable for design thinking processes, but the unclarity leads to unmet expectations. We propose to clarify what is meant by the word 'prototype', reserving it solely for representations that portray what realistically can be and use 'idealtype' for representations that portray what is preferred: the ideal.

Introduction

'I love it! When will you have a prototype?', the sponsor of an assignment of the Urban Leisure and Tourism (ULT) lab asked the students. She had just witnessed and experienced a bold presentation for a story-sharing tree (Figure 1), a design intervention imagined by an interdisciplinary group of students working in a lab, an experimental learning environment. The concept suggested a place where visitors can sit around a tree full of lights at night to share stories. Once they are finished, they add another light. The students carefully prepared their presentation and acted out their concept with several beneficiaries, who had to sit on the ground around a small tree full of lights and share a personal story. It was a great experience, and everybody present seemed to appreciate the concept. So, the partners asked when they would have a prototype, the students were taken aback. In

their view, they just presented a prototype: the small tree full of lights! Worse: it was the final presentation, so what else did the partner expect? They looked at their coaches for help, who asked what the assignment sponsor meant, as what was before her was the prototype. Now she was confused: *'This? This not a prototype, right?'*

Figure 1. The prototype of a story-sharing tree to enhance the perceived safety in a neighbourhood.

For the authors, this was not the first time that different expectations of a 'prototype' were manifested and revealed. It is one of those terms deeply ingrained into the vocabulary of design thinking, but what is meant by it? Practitioners use it often, sometimes even as a verb (prototyping) but seldom offer an explanation of what they mean by it. They consider it a term that is clear to everyone, which needs no further explanation. Likewise, in popular textbooks on design thinking (e.g., Brown, 2008; Den Dekker, 2019; Martin, 2008; Stompff, 2018), the word is omnipresent, but seldom a concise definition is offered. Effectively, the word is used for all kinds of representations, ranging from sketches on the back of a napkin to elaborate demonstrations or functional models of high-tech products. These prototypes are not the same and serve different goals in the design process. By being implicit about what is meant by a prototype (and what goal it serves), novices engaged with design thinking hardly have a clue why they must produce prototypes, or what standards those prototypes need to meet to be effective.

Setting the Stage

With the proliferation of design-infused methods to problem solving, it is not just expert designers who involve themselves with design. Simon (1996, p.111) argues that 'everyone designs who devises courses of action aimed at changing existing situations into preferred ones.' This argument is of relevance in the context of experimental environments, where many new ideas are developed and tested by people who are not trained and/or working as a designer.

Over recent decades, numerous bestselling books have popularised the concept of design thinking, and design thinking has permeated various sectors, including healthcare and social services (Brown, 2008; Liedtka & Ogilvie, 2011; Martin, 2008). Today, the process is embedded in numerous university curricula (Wrigley & Mosely, 2022). Within the discourse on design thinking, however, starkly different schools of thought exist that attribute different meanings, depending on the context in which they operate (Johansson Sköldberg et al., 2013). In line with Simon's (1996) original position, we consider design thinking as an approach not left to design experts alone: everybody designs or needs to design every now and then. Yet, even though 'everybody is endowed with the ability to design, (..) not everybody is a competent designer and few become professional designers' (Manzini, 2015, p. 37). We frame design thinking as a way to democratise design, to empower non-designers to design and/or participate in a design process. First, by teaching designerly ways of thinking to all kinds of professionals (Stompff, 2022). And second, by actively involving stakeholders in the design process, learning-by-doing, and designing together. Labs serve as playgrounds where non-designers can (learn to) design, together with designers or coaches with design expertise, explicitly crossing disciplinary boundaries, a theme of this book.

Yet, when design thinking became widely embraced, design theorists such as Dorst (2011) and Kimbell (2011) raised concerns that crucial aspects of design practice are overlooked by scholars. The latter argued that thinking and making are disconnected and the role of design artefacts are underestimated (Kimbell, 2011). Designers sketch, make, create, and test all kinds of communicable representations. To enhance the theory of design thinking, it is necessary to give scholarly attention to the construction of these representations, as part of the design process – and to develop a common language for analysis and comparison. This chapter focuses on a specific kind of representation that is omnipresent in design practice and in books on design thinking: prototypes. Two research questions guide the study:

- *What does a prototype represent?*
- *What goals does it serve in design thinking processes?*

We first offer a literature review on what prototypes are and what goal they serve. An analytical framework is presented that offers a lens with which to study prototypes on the basis of what they represent. After setting out the methodological approach, findings are presented, which illustrate a fundamental ambiguity that leads to unmet expectations of beneficiaries and other stakeholders. Lastly, we put forth a recommendation for a more precise vocabulary to enhance design thinking processes and education.

Literature Review: Representations

Numerous schools of thought can be observed in discussions on design thinking, originating either from the managerial realm or from design sciences (Johansson Sköldberg et al., 2013). The lack of clarity and the 'more superficial and popular character' of the management discourse (Johansson Skolberg et al., 2013, p.121) places the paradigm at risk of 'construct collapse' (Micheli et al., 2019, p. 124), whereby the promise it suggests is never met. Surprisingly, these different schools of thought seldom discuss the relationship between '*design*' and '*thinking*', or more precisely, between the activities of designing and thinking. By not addressing this question, the underlying epistemology remains unclear, and a knowledge gap exposes around how knowledge is created through design (Dixon, 2020; Rylander Eklund et al., 2021).

Two paradigms on knowledge creation through design

A widely embraced approach for design thinking is the Double Diamond process of the British Design Council, which is central to acclaimed books on design thinking (Boeijen et al., 2020; Den Dekker, 2019; Lewrick et al., 2018). It explicitly divides the process into two distinct phases: the first 'diamond' revolving around discovering and defining a problem, and the second around developing and delivering solutions. Problem setting precedes the more creative parts, effectively separating analysis from creation. Likewise, making and testing things is named relatively late in the process, implicitly separating thinking from making. Prototypes are considered the result of a careful thinking process.

A well-known paradigm surfaces that can be characterised as *thinking precedes making*: what designers create, such as prototypes, are artefacts of what they considered and thought before. It is reminiscent of the Cartesian duality that separates mind and body, the mental and the physical. What individuals do is coordinated by what they think a priori what they do: thinking precedes action.

Within the domain of design sciences, an alternative paradigm exists and informs significant contributions to the understanding of design cognition (Cross, 1982; Dorst, 2011; Schön, 1983; 1992; Sennett, 2008; Visser, 2006). This paradigm rejects the duality of thinking and doing and considers what we think is mediated by what we do and vice versa. It can well be clarified through Schön's (1983) concept of 'reflection-in-action', whereby a designer thinks through iteratively creating and reflecting on outcomes. For example, whilst sketching a designer reflects on the evolving design and responds by adjusting the sketch in realtime or by creating new ones. This process represents a dynamic 'conversation with the situation', that unfolds through sketching (Schön, 1992). Remove technology, pencil and paper, and a designer can no longer design. In this view, *thinking and making are mutually constitutive*, responding to each other akin to a dance (Stompff et al., 2022a). What is designed shapes what designers think, and what designers think shapes the emerging design. This is not different for a (co)production process whereby feedback loops inform the design process (Horgan & Dimitrijević, 2020).

Representations and prototypes

Visser, a prominent design theorist, put the lens on the artefacts of design themselves. She considered that the construction of representations is the core of design cognition (Visser, 2006). Although the form of these representations may vary depending on what is designed, the ongoing construction of representations is what characterises design (Visser, 2009). Designers advance their ideas by ongoingly creating and adapting representations.

To bypass a lengthy and philosophical discussion about representations and what they signify, we provide a broad definition: *a representation is a portrayal of something else in some form*, which can take the shape of a drawing, model, text, diagram, or blueprint for assembly.

A representation can portray something from the 'real' world, like an icon symbolising a car, or it can portray a mental construct, such as an idea. In design, a representation mostly concerns a portrayal of an idea or concept in some form, which enables the designer and others to reflect on the idea. In this way, the representations are 'in flux, continuously adapted and never complete' (Stompff & Smulders, 2015). In time a design arises whereby every detail is described by means of representations of some kind, ranging from technical drawings up to tangible prototypes.

These representations vary significantly in their fidelity, which is the degree to which the representations correspond to what can be tested within a 'real world' scenario. High-fidelity representations closely resemble the 'real' world, whereas low-fidelity representations do not necessarily look like the actual things they represent, albeit their meaning is clearly conveyed, much like how a person is depicted on a traffic sign. The fidelity of representations and prototypes is a well-debated topic, for example in the fields of user experience design (Rudd et al., 1996; Sauer & Sonderegger, 2009), or design in team and company settings (Stompff & Smulders, 2015; Lauff et al., 2018).

Prototype as a representation of an idea

In both literature and design practice, the term 'prototype' is frequently used, yet its precise meaning can vary considerably. Some scholars adopt a broad perspective as exemplified by Houde & Hill (1997) in their influential work. They argue that prototypes encompass any form of representation of a designer's idea, regardless of the medium used. Consequently, sketches can be a prototype, just as a small movie can be a prototype (Houde & Hill, 1997, p.368). The ULT lab students considered their presentation (Figure 1) to be a prototype of their proposed intervention.

As this meaning of a 'prototype' is broad, scholars developed various taxonomies to classify different types of prototypes that represent an idea. This includes distinctions between form and function (Hallgrimsson, 2012), virtual and physical prototypes

(Sefelin et al., 2003), and low and high-fidelity prototypes (Lim et al., 2008; Rudd et al., 1996). Other scholars discuss prototypes when they are created in the design process (Camburn et al., 2017), distinguishing for example *'pretotypes'* as prototypes made in the fuzzy front end (Savoia, 2011).

A popular approach to discuss kinds of prototypes is to focus on the medium used, such as 'paper prototypes' (Snyder, 2003), 'sketch prototypes' (Buxton, 2010), or 'virtual reality prototypes' (e.g., Nebeling et al., 2019). Echoing McLuhan's (1967) famous maxim, *'the medium is the message'*, the inclusion of the medium in these discussions has a reason. The different materialities of prototypes offer different affordances and thus serve other goals. For example, paper prototypes are cheap and extremely easy to create even by non-designers.

Prototype to validate assumptions

Some experts have a more confined view on prototypes, just like what the assignment sponsor considered when listening to the presentation of the lab students (Figure 1). They consider a prototype a representation of the eventual outcome of the design process, made with the purpose of answering a question or testing an assumption (Lauff et al., 2019; Menold et al., 2018). Expert designers and design teams advance their design process by conducting numerous tests with many prototypes throughout a project's lifecycle. It helps them to discern which ideas are effective or not. This continual process of testing and refining ideas shapes the characteristic cyclical nature of design, what Lake (2021) and colleagues refer to as *'iterative prototyping'*.

According to this perspective, the primary focus shifts away from the inherent properties of the prototype to its performance in achieving a predetermined goal, such as user-friendliness, mechanical strength, or market acceptance. As Gaver (2014) argued, the accountability of design is that it must work. Thus, prototypes are created purposefully, with the aim to conduct specific tests, such as a drop test or a usability test, which validate whether the design works as expected. The test goals result in dedicated terms, such as 'usability prototypes' (Hall, 2001) or 'experience prototypes' (Buchenau & Suri, 2001).

Prototype as vehicle of (collaborative) design

Prototypes are also useful as artefacts to facilitate collaborative design, and co-learning where domain-shift is required among a diverse group of problem-solvers. In this way, prototypes become performative, where the act of creating prototypes takes centre stage in advancing the design process. This is particularly relevant in the context of interdisciplinary design teams (Bucciarelli, 1994; Stompff & Smulders, 2015) and product-service design (Kleinsmann & ten Bhömer, 2020). Likewise, prototypes serve also as tools in organisations to enhance communication, facilitate learning, and support informed decision-making (Lauff et al., 2018). Expanding beyond organisational boundaries, the practice of prototyping can also drive design research (Wensveen & Matthews, 2014) and facilitate co-design processes of users and researchers (Bødker & Grønbæk, 2022).

In this perspective, prototypes are seen as *both artefacts and enablers of thinking*, as nascent ideas are brought to life through prototyping and experimentation. It is in line with the paradigm whereby thinking and making are mutually constitutive, neatly summarised as *'building to think'* (Brown, 2009, p.87). Designers use foam blocks and duct tape to create a quick, tangible representation of an idea to offer a better understanding of its dimensions and characteristics. Often prototypes disclose surprises that enable design teams to learn and innovate (Stompff & Smulders, 2015). Prototypes are intermediate outcomes of a collective cognitive process aimed at creating something new, a future reality that is yet to come into existence. Here, the emphasis is placed on the process itself, and the transdisciplinary combination of knowledge, domains, and perspectives (Horgan & Dimitrijević, 2020).

Prototypes and design thinking education

The two contrasting paradigms on knowledge creation through design significantly influence design thinking education and the role of prototypes within it. When programs are set up with the assumption that thinking precedes making, students are encouraged to follow a structured path, for example deploying the Double Diamond approach. In this way, research activities and analysis are mainly conducted before generative design activities as ideation are done. The insights gained from research inform the subsequent design process, ultimately leading to the creation of a prototype.

On the other hand, if educational programs assume that design thinking and design doing are inseparable – parallel and mutually responsive – students are encouraged to engage a *'build to think'* mindset. They create prototypes often, starting early in their project, as the prototypes inform the design process. Students engage in iterative prototyping (Lake et al., 2021), whereby the artefacts and the feedback on those artefacts inform research (Gaver, 2012; Zimmerman et al., 2010). A prototype is thus both an artefact and a driver of the design process.

Wrigley & Mosely (2022) discussed many educational programs at universities that address design thinking, each suiting their specific objectives and student populations. However, as Lake et al. (2021, p. 349) observed, these educational programs 'only rarely required students to engage in iterative prototyping'. Put differently, most programs embrace the paradigm whereby thinking precedes making and teach students to conduct research, design and eventually build a prototype.

Analytical Framework

In short, prototypes can be classified in different ways, such as on the basis of their material properties, their fidelity, or their goal. However, what remains undertheorized is *what a prototype represents and how that representation impacts design thinking processes*. A framework for analysis (Figure 2) has been developed, drawing inspiration from Nelson & Stolterman's (2014) influential model of 'the design way', describing design as a pathway of action and inquiry for anyone. Nelson & Stolterman (2014, pp. 27-40) argued that designers grapple with three interrelated inquiries, which they have termed inquiries into 'the truth', 'the ideal', and 'the real'. Each inquiry produces different kinds of knowledge (Van Turnhout et al., 2019; Turnhout & Smits, 2022): the first inquiry produces knowledge that explains and describes the current situation, the second inquiry offers insights into what is aspired and desired, and the third results in knowledge that prescribes and predicts what effective mechanisms and solutions are.

To avoid getting entangled in complex philosophical debates regarding the nature of 'truth' and 'reality', we believe we need another vocabulary closer to design practice and Simon's (1996) clear vocabulary, who considered design as devising plans of action to change an existing situation into a more preferred one. Accordingly, design encompasses three interrelated inquiries, guided by three key questions:

- What is the existing situation?

This question delves into the current situation, seeking to understand its underlying causes and effects to set the problem at hand. It is an inquiry into what stakeholders believe the problem is.

- What is the preferred situation?

This question explores what stakeholders want or desire: the ideal. It sets the proverbial dot on the horizon collectively, shaping the values guiding the design process. This question does not solely concern goals but offers a vision of how to obtain these goals by means of inspiring conceptual ideas. It is an inquiry into what matters for stakeholders.

- What can be?

This question revolves around the plans of action, where realistic goals must be balanced with practical considerations such as manageable budget, time, and available resources. It is an inquiry into what works adequately for the stakeholders.

In line with Visser (2006, 2009), we consider that in order to advance these three inquiries, dedicated representations are needed. The three interrelated inquiries are different, and the question is what representations advance each inquiry. The framework provides a lens to study prototypes in an innovative way, to consider what kind of representations and prototypes serve which inquiry.

Method: Research Through Design

The chosen method is a *research through design* approach, resembling action research experiments and observation. It takes a decisively pragmatist stance, using Dewey's seminal ideas on research as a start (Dixon, 2020; Stompff et al., 2022a). In this approach not only design methods and outcomes are a legitimate part of research (Gaver, 2012; Zimmerman et al., 2010), but also conducting experiments in practice are tenet. These experiments offer guidance to (1) develop transferable knowledge; (2) by oscillating between known theories and uncharted practices until new ideas arise; whereby (3) the value of these ideas is validated through experiments. Such a study develops not only a solution to a problem, but also novel, explicit, transferrable, and accumulable knowledge.

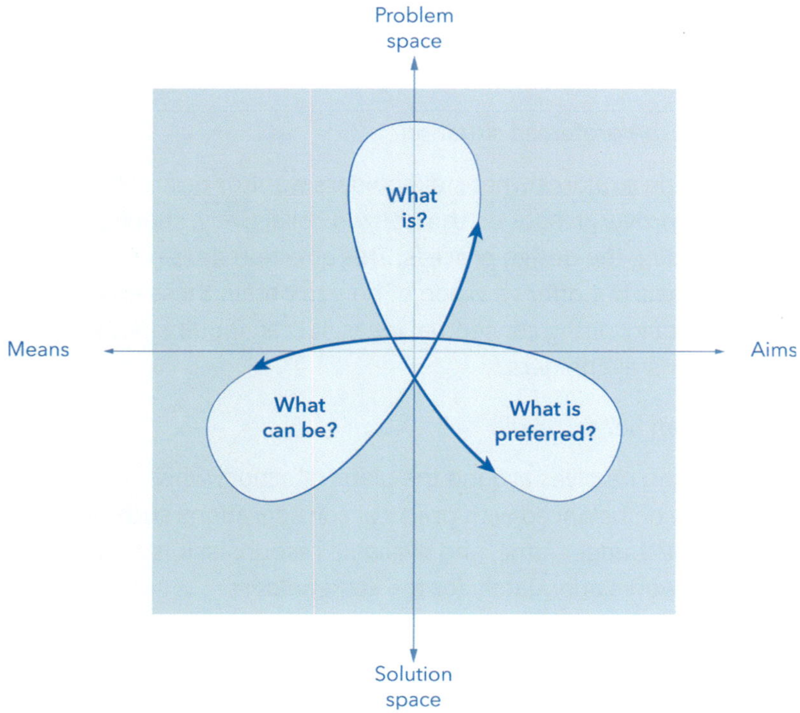

Figure 2. Framework for Analysis. Design comprises three interrelated inquiries that are mutually constitutive: examining what the existing situation is, what the preferred situation is, and what can be, that is, what feasible and viable plans of actions are. The three inquiries are positioned against two dimensions. Whereas the 'What is?' question explicitly explores the problem space, the two other inquiries explore the solution space. The inquiry into 'What can?' focuses more on means, whereas 'What is preferred?' has a focus on goal-setting.

A study (Figure 3) is initiated by a researcher as a result of an unexpected event causing doubt, requiring setting the problem provisionally. An iterative design process first begins with *making sense* of the current situation, interrogating what is causing the problem at hand by comparing practical observations with known theories. In subsequent stages, *logical reasoning* starts to decide what can be done and what anticipated outcomes are. Effectively, these are hypotheses in a particular format: if X is done, then Y is expected to happen. These hypotheses can be tested *through experimentation* in practice. If outcomes of these tests do not yield anticipated outcomes, another iteration starts, to make sense, reason, and experiment once again. A problematic situation is adequately understood once outcomes of interventions are in line with expectations. In this approach, researchers can only present an argument and evidence in hindsight, once an adequate solution is found: *'(i)n this way and only in this way, do we act logically'* (Dixon & French, 2020, pp.19-20).

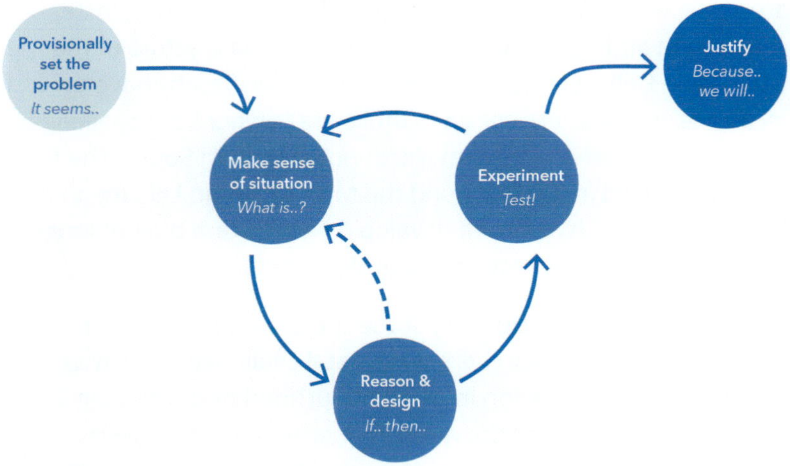

Figure 3. Overview of a research through design in practice, consisting of several steps, whereby in an iterative way and through experimenting new knowledge is produced.

The research setting is within an interdisciplinary educational context (Living Lab) that is open for experimentation and whereby design thinking is at the heart of the program. In a period of more than 2 years, the researchers engaged with hundreds of students and approximately thirty coaches. First, they made sense of the problem at hand, followed by offering informed masterclasses on prototyping and reflecting on what students created. The format of the masterclasses was improved upon several times before students began to produce prototypes that more closely aligned expectations of stakeholders and advanced the projects and learning outcomes. New theory was developed, and the resulting intervention tested in practice.

The Urban Living Labs of Inholland University of Applied Sciences

The design inquiry took place within a University of Applied Sciences, mostly in six interdisciplinary Urban Living Labs (ULLs) spread over several locations in the Netherlands. Living Labs are real-life test and experimentation environments that foster co-creation and open innovation among the main actors of the Quadruple Helix (ENoLL, undated). These specific labs have a local presence and are set up in those areas where the challenges manifest, such as Amsterdam North or Rotterdam South. The first ULL was started in 2015, around the topic of Urban Leisure and Tourism, and all ULLs aim to develop an ecosystem built of long-term relations with a network of local partners.

In these ULLs, interdisciplinary student teams work together with coaches and researchers on societal challenges – such as enhancing social cohesion in a neighbourhood or developing a healthier ecosystem for the music industry – over a semester-long lab track. Undergraduate students from across several study programmes participate (e.g., tourism management, business innovation, and communication), working on assignments set by project sponsors, and closely with local communities and end-beneficiaries. The students must make sense of the situation, develop new innovative frames, design concepts, and test prototypes towards real-world implementation.

Figure 4. ULL of Inholland University of Applied Sciences, located in Amsterdam North.

Findings: Shaping the Ideal and the Real

Strikingly consistent patterns of related problems emerged across various labs and over time. These patterns recurred each time new students enrolled in the educational program or when new coaches joined. Below, we present the three key findings and their corresponding interventions that mitigate the problems at hand.

Reluctance to be expressive

An idea is a thought or suggestion as to a possible course of action. It exists in the mind and to get feedback or to test the idea, it needs a representation allowing the individual who has the idea or others to reflect on it. Put differently, an idea must be expressed in some format, which can be a story, sketch, model, movie, and so on. The representations allow us to reflect and inspire.

One of the most persistent findings in our research is the reluctance of non-designers to express their ideas visually or physically, even when they are explicitly encouraged to do so. They often lack confidence and tend to rely on verbal and textual descriptions instead. Yet, text is often not the best form to express an idea: it is hard to describe how a product looks, how a user interface is organised ,or to experience an idea for a new service. Only after investing significant time in developing tangible representations are students willing to share it. Even then they openly display feelings of discomfort: creating representations is a burden for them. The implication is that non-design students only produce visual and tangible representations once they developed a clear idea in their minds, rather than producing representations to develop an idea!

We believe this tendency is not a deliberate choice but is heavily influenced by a lack of expressive skills. They did not receive formal training in sketching or modelmaking, and students (and coaches) rely on verbal and written argumentation until they are certain about their design direction. Unfortunately, this tendency implies that ideas are not always fully represented well. This can prevent other stakeholders from understanding the idea, and its potential impact, to offer feedback or contribute. To address this issue, we have found that the establishment of a *'show and share'* culture helps, whereby students regularly display their work to each other, creating a more open and supportive learning environment, and a culture that encourages the use of tangible representations (Stompff et al., 2022b).

Preoccupation with the model, the ideal

Using the framework (Figure 2) as a lens, it became evident that both coaches and students were unaware of the difference between design inquiries into what should be (the preferred situation) and what can be (implementation roadmap, plans of action). Consequently, it was unclear what the representations produced were referring to, including those labelled as *'prototypes'*. Are the prototypes representing ideas on that which should be, or are they representing the eventual design, that which can be? The lack of a well-considered taxonomy relating the prototype to the context in which a proposition is being interrogated (tested) is a clear impediment to advancing a design thinking process. The case as described in the introduction demonstrates this ambiguity well: students made a prototype, which represented an idealistic and inspiring idea, yet the partner was hoping for a realistic prototype considering the limited available budget. The ambiguity in what the prototype represents directly shapes the expectations regarding what the prototype seeks to conceptually test or communicate.

We observed that novices to design thinking often assume that prototypes refer to representations of the ideal (what should be): most of the prototypes made by students clearly embody the ideal (Figure 5). Far fewer prototypes produced by students portray a realistic design of a product, app, or plan of action (Figure 6). Yet different kinds of representations (including prototypes) serve different inquiries of the design process. The ambiguity effectively leads to miscommunication and unmet expectations among stakeholders.

The prototypes that serve the different inquiries have specific characteristics and in Table 1 an overview is provided.

Table 1. Specific characteristics of prototypes.

	Prototypes that represent the ideal	**Prototypes that represent realistic plans of action**
Aim	Advancing the inquiry into what should be, with stake-holders	Advancing the inquiry into what can be, with stake-holders
Fidelity	A low fidelity, embodying a narrative rather than out-come	A medium or high fidelity, depicting accurately what is intended
Nature or pur-pose	Conceptual, speculative, exploring an idea without caring too much about feasi-bility and viability	Refined, realistic, often func-tional, allowing to conduct tests and validate design choices
Openness for interpretation	Open, depicting a vision but leaving open how to make it happen	Closed: what you see is what you (will) get

Representations of the ideal provide compelling vision to aspire to, and they play a crucial role in facilitating discussions among stakeholders, coaches, and other students to define the project's goals. They advance a collective consensus into what should be, as these representations enable stakeholders to establish a shared understanding of what is considered 'good' or valuable. These representations contribute positively to the design (thinking) process, especially when various conflicting interests are at play. The representations help to initiate conversations about what constitutes 'good' for all involved, hopefully transcending conflicting stakes.

However, these 'prototypes' do have their limitations. They omit challenging design decisions that must be made to create feasible and viable solutions – often related to funding, ongoing management, and operationality. The true difficulty of design often lies in making choices, implicating that certain goals cannot be fully realized due to constraints like budget limitations. Also, a prototype representing the ideal lacks the practicality needed to conduct real-world tests, such as a usability test. Consequently, stakeholders may perceive these prototypes made by students as representations of wishful thinking, devoid of realism – and can dissuade project sponsors from committing to a long relationship with the Living Lab.

Figure 5. Some prototypes made by student teams in the ULLs that represent the ideal, advancing the inquiry into what should be. On the bottom, a miniature diorama is shown that depicts what should be from a specific angle.

Representations that portray realistic designs or plans of action advance the inquiry into what can be. Creating these prototypes helps to make decisive choices, spanning from high-level conceptual choices to intricate details that each impact the functionality, cost, and limitations of the design. In this way, an aspirational dream slowly transforms into a more realistic design, leaving little room for interpretation. These prototypes hold significant value in the design thinking process primarily to validate whether a design works or not, meaning it meets the established goals and standards. They hold important currency for informed decision-making on future policy innovation and the success of socially innovative interventions (Horgan & Dimitrijević, 2018).

However, as students discovered, presenting realistic prototypes of 'that which can be' can also lead to disappointment among stakeholders. They hope for more ambitious plans than what is presented to them. On multiple occasions, coaches found themselves obliged to step in between students and stakeholders as the latter expressed disappointment. Balancing the desire for a compelling vision with the practical realities of implementation can be a challenge in design processes.

One of the big issues with design thinking education is that students lack relevant expertise to craft realistic prototypes that can be put to test. Process interventions cannot fully repair this issue, and easily accessible expertise is needed to enhance design thinking processes. Learning by doing remains the most powerful way to disseminate these methods.

Figure 6. Some prototypes made by students that represent what can be. These prototypes project the intended outcome and are tested in practice. The photo at the bottom shows students (with rain ponchos) who made a wishing wall, and whilst making it already the first passersby joined.

Difficulty to iterate

Observations and reflections suggest that iterative prototyping does not come naturally to students lacking design expertise. Likely driven by a restraint to work visually, the creating of prototypes often stalls until the very last moment. Consequently, tests are also conducted late – if they are conducted at all – and students inevitably discover that the tests work out differently than expected, leaving them little time to improve their concepts.

Implicitly, they are encouraged by coaches (in cases lacking relevant design expertise), who sometimes see prototypes as the 'end product' of design thinking rather than a vehicle for idea development.

Through interventions in the program, students were obliged to engage in iterative prototyping and robust *testing*. They were encouraged to begin to develop prototypes in the early stages of their projects and to 'go outside', test and fail fast. In this way, students learned that a mindset of prototyping and testing ideas builds confidence, as it becomes clear what works and what does not. For instance, students working on the issue of littering in an urban context believed that an interactive garbage bin could help address the problem. They placed a speaker inside a garbage bin in a public square and interacted with passersby from a distance, encouraging them to dispose of their waste in the talking bin rather than on the street. The results of this preliminary prototype test demonstrated that the idea was effective, allowing them to continue to experiment with various interaction strategies.

Interestingly, although many students find it initially challenging (and embarrassing) to test their half-baked ideas, they quickly embrace the mindset of learning through experimentation. They realise that testing ideas in practice provided a solid foundation for justifying decisions, not only to assessors but also to critical stakeholders, end-users, and beneficiaries.

Discussion: the Need for Disambiguation

This study demonstrates the lack of clarity surrounding the term 'prototype' and what it represents to those involved in the design process, and the resultant confusion. One type of prototype depicts the ideal and advances the inquiry into what should be, by helping to build consensus among stakeholders *on what matters*. These prototypes embody a vision that goes beyond abstract goals as 'sustainable' and transcend contradictory stakes by projecting a meaningful direction in which the eventual solution might be found. For this inquiry into what should be, *consensus among stakeholders on what matters justifies made design choices* and is less about feasibility and viability. This inquiry is echoed in recent books on social design whereby a focus is on meaning-making activities such as storytelling or role-playing (Amatullo et al., 2021; Manzini, 2015).

Prototypes that correspond to concrete plans of action advance the inquiry into what can be. Put differently, how do we get where we want? These representations, whether they pertain to a product, app, service, or another intervention, are the synthesis of numerous design choices and can be tested to demonstrate that predefined goals, standards, and criteria are met. The representations help to settle *what works* in practice and the *validation in practice justifies made design choices*. In this inquiry, practical consequences take precedence, and opinions are of lesser importance.

Recognizing the distinct kinds of representations and the different inquiries they serve can lead to more effective communication and collaboration among stakeholders and designers. The ambiguity of the word 'prototype' leads to confusion and unmet expectations. Therefore, we propose introducing a new term. The term 'prototype' originally derives from Greek, where *'protos'* means 'first', and *'typus'* refers to 'kind or type.' Historically, a prototype was the first form that served as the basis for creating something new to be reproduced. With this historical context in mind, we therefore suggest reserving the term *'prototype'* for representations that correspond to the eventual outcome, advancing the inquiry into what can be.

In contrast, we propose using the term *'idealtype'* to refer to representations that correspond to the ideal, advancing the inquiry into what should be. Idealtypes serve the purpose of reaching consensus and a shared understanding on what is valued by stakeholders, offering a vision to guide subsequent design activities.

This distinction in terminology is aimed at enhancing communication and understanding among stakeholders, towards a better alignment of expectations. Each inquiry demands specific representations and the proposed distinction between *prototypes* and *idealtypes* facilitates discussions with stakeholders on what matters and on what works separately. The relevance of this distinction is particularly evident in social and socially driven design contexts, where stakeholders often have diverse and potentially conflicting perspectives and values. In such situations, *idealtypes* can serve as representations that embody visions and ideals, fostering discussions and consensus-building around what matters most and what should be considered good.

Recommendations

Although no comparisons are made, the confusion on what representations are seems to strengthen in these specific experimental environments. Individuals lacking design expertise who partake in collaborative design often struggle to articulate their thoughts and rarely resort to interactive prototyping. Additionally, a preoccupation with the ideal becomes evident, overlooking the necessity of practical testing to determine whether ideas align with expectations. These tendencies among participants in experimental environments impede the creative design process.

Although it doesn't address the issue of expressive skills deficiency, introducing an adapted vocabulary can enhance the management of expectations. By incorporating terms such as *idealtypes* and *prototypes* and dedicating time to develop and present both, it enables focused inquiries into what is preferred and what can be. This approach serves specific goals within the design process, fostering a more effective exploration of what is desired and what can be feasibly realised.

The study is chiefly concerned with reflections from the Urban Living Lab setting, from interactions with students who, for the most part, lack design expertise. If we are to consider design thinking the democratisation of design, then consequently the setting of design novices working together with external stakeholders offers a good environment for conducting experiments to improve design collaboration. However, this simultaneously severely limits the generalisability of our findings, for which we wish to continue this study outside the specific context – and to develop a keener understanding of the use of the terms within other design thinking settings.

The investigation may therefore advance on two frontiers. First, we need to delve into the third inquiry of Nelson & Stolterman's (2014) framework that probes into the existing, problematic situation. Which representations advance this inquiry with stakeholders, besides considering visual maps, infographics, or data visualisation methods? An interesting concept in this context are '*provotypes*' (Boer & Donovan, 2012), or representations that explicitly are aimed to provoke deep discussions among stakeholders to unravel underlying assumptions.

Second, there is an extensive body of literature on boundary objects, objects that are *'both plastic enough to adapt to local needs and the constraints of the several parties employing them, yet robust enough to maintain a common identity'* (Star & Griesemer, 1989, p. 393). Put differently, objects that enable collaboration and span boundaries across practices that find it hard to understand each other (see also Chapter 8, The Open Lab as Boundary Object). Prototypes serve as *boundary objects,* and this concept may provide us with a framework with which to better understand the contribution of prototypes to participative design.

Figure 7. Streetwork depicting ReneMagritte's famous work *The Treachery of Images*, source Wikimedia Commons.

Conclusion

This study calls to mind the painting *The Treachery of Images* (Figure 7) by Magritte. This surrealist masterpiece shows a pipe and the text 'Ceci n'est pas une pipe', questioning what is meant with a representation of pipe. Likewise, if someone names something a 'prototype' what does this person mean? The findings shed new light on representations and meaning, transcending the specific context and contributing to the ongoing discourse on design education and practice. The distinction between *prototypes* and *idealtypes* provides a practical framework for navigating the complexities of design – above all in social contexts, where diverse perspectives and values are at play. Although the question remains whether these two words are the best to disambiguate representations, the distinction explicates and advances two interrelated inquiries: an inquiry into *what matters to those involved*, and an inquiry into what can be, that is, *what works in practice*. Both types of inquiry are relevant for design and design thinking processes, yet most stakeholders are unaware of their subtle yet characteristic differences. By using two terms, *idealtypes* and *prototypes*, the process becomes better articulated and understood. This allows stakeholders to engage meaningfully by reflecting, commenting, and contributing to the ideas represented.

Bibliography

- Amatullo, M., Boyer, B., May, J., & Shea, A. (Eds.). (2021). *Design for social innovation: Case studies from around the world*. Routledge.
- Bødker, S., & Grønbæk, K. (2020). Design in action: From prototyping by demonstration to cooperative prototyping. In Grudin, J., Bardzell, S., Bardzell, J., Bødker, S., Boy, G. A., DiSalvo, C., & Pargman, D. (Eds.), *Design at work* (pp. 197-218). CRC Press.
- Boer, L., & Donovan, J. (2012, June). Provotypes for participatory innovation. In Proceedings of the designing interactive systems conference (pp. 388-397).
- Boeijen, A. G. van, Daalhuizen, J., & Zijlstra, J. (2020). *Delft design guide: Perspectives, models, approaches, methods*. BIS Publishers.
- Brown, T. (2008). Design thinking. *Harvard Business Review, 86*(6), 84.
- Bucciarelli, L. L. (1994). *Designing engineers*. MIT Press.

- Buxton, B. (2010). *Sketching user experiences: getting the design right and the right design*. Morgan Kaufmann.
- Camburn B., Viswanathan, V., Linsey, J., Anderson, D., Jensen, D., Crawford, R., & Wood, K. (2017). Design prototyping methods: state of the art in strategies, techniques, and guidelines. *Design Science, 3*, e13.
- Cross, N. (1982). Designerly ways of knowing. *Design studies, 3*(4), 221-227.
- Dekker, T. den (2020). *Design thinking*. Routledge.
- Dixon, B. S. (2020). *Dewey and Design: A Pragmatist Perspective for Design Research*. Springer Nature.
- Dixon, B. S., & French, T. (2020). Processing the method: Linking Deweyan logic and design-in-research. In *Design Studies, 70*, 1-23.
- Dorst, K. (2011). The core of 'design thinking' and its application. *Design studies, 32*(6), 521-532.
- ENoLL, European Network of Living Labs (undated). Retrieved from https://enoll.org/about-us/
- Gaver, W. (2012). What should we expect from research through design? In Proceedings of the SIGCHI conference on human factors in computing systems (pp. 937-946).
- Gaver, W. (2014). Science and Design: The implications of different forms of accountability. In Olson, J., & Kellogg, W. (Eds.), *Ways of Knowing in HCI* (pp. 143-165). Springer.
- Hall, R. R. (2001). Prototyping for usability of new technology. *International Journal of Human-Computer Studies, 55*(4), 485-501.
- Hallgrimsson, B. (2012). *Prototyping and modelmaking for product design*. Hachette UK.
- Horgan, D., & Dimitrijević, B. (2020). Social innovation in the built environment: the challenges presented by the politics of space. *Urban Science, 5*(1), 1.
- Houde, S., & Hill, C. (1997). What do prototypes prototype? In Helander, M. (Ed.), *Handbook of human-computer interaction* (pp. 367-381). North-Holland.
- Johansson Sköldberg, U., Woodilla, J., & Çetinkaya, M. (2013). Design thinking: past, present and possible futures. *Creativity and innovation management, 22*(2), 121-146.
- Kimbell, L. (2011). Rethinking design thinking: Part I. *Design and culture, 3*(3), 285-306.

- Kleinsmann, M., & ten Bhömer, M. (2020). The (new) roles of prototypes during the co-development of digital product service systems. *International Journal of Design, 14*(1), 65-79.
- Lauff, C.A., Kotys-Schwartz, D., & Rentschler, M.E. (2018). What is a Prototype? What are the Roles of Prototypes in Companies? *Journal of Mechanical Design, 140*(6), 061102.
- Lauff, C., Menold, J., & Wood, K.L. (2019). Prototyping Canvas: Design Tool for Planning Purposeful Prototypes. In *Proceedings of the 22nd International Conference on Engineering Design (ICED19)*. DOI:10.1017/dsi.2019.162
- Lake, D., Flannery, K., & Kearns, M. (2021). *A cross-disciplines and cross-sector mixed-methods examination of design thinking practices and outcome. Innovative Higher Education*, 46(3), 337-356.
- Lewrick, M., Link, P., & Leifer, L. (2018). *The design thinking playbook: Mindful digital transformation of teams, products, services, businesses and ecosystems*. John Wiley & Sons.
- Liedtka, J., & Ogilvie, T. (2011). *Designing for growth: A design thinking tool kit for managers*. Columbia University Press.
- Lim, Y. K., Stolterman, E., & Tenenberg, J. (2008). The anatomy of prototypes: Prototypes as filters, prototypes as manifestations of design ideas. *ACM Transactions on Computer-Human Interaction (TOCHI), 15*(2), 1-27.
- Manzini, E. (2015). *Design, when everybody designs: An introduction to design for social innovation*. MIT Press.
- Martin, R. L. (2009). *The design of business: Why design thinking is the next competitive advantage*. Harvard Business Press.
- Menold, J., Simpson, T.W., & Jablokow, K. (2018). The prototype for X framework: exploring the effects of a structured prototyping framework on functional prototypes. *Research in Engineering Design*. doi.org/10.1007/s00163-018-0290-9.
- Micheli, P., Wilner, S. J., Bhatti, S. H., Mura, M., & Beverland, M. B. (2019). Doing design thinking: Conceptual review, synthesis, and research agenda. *Journal of Product Innovation Management, 36*(2), 124-148.
- Nebeling, M., & Madier, K. (2019). 360proto: Making interactive virtual reality & augmented reality prototypes from paper. In Proceedings of the 2019 CHI Conference on Human Factors in Computing Systems (pp. 1-13).

- Nelson, H. G., & Stolterman, E. (2014). *The design way: Intentional change in an unpredictable world*. MIT Press.
- Rudd, J., Stern, K., & Isensee, S. (1996). Low vs. high-fidelity prototyping debate. *Interactions, 3*(1), 76-85.
- Rylander Eklund, A., Navarro Aguiar, U., & Amacker, A. (2021). Design thinking as sensemaking—Developing a pragmatist theory of practice to (re) introduce sensibility. *Journal of Product Innovation Management.*
- Sauer, J., & Sonderegger, A. (2009). The influence of prototype fidelity and aesthetics of design in usability tests: Effects on user behaviour, subjective evaluation and emotion. *Applied ergonomics, 40*(4), 670-677.
- Savoia, A. (2011). Pretotype it. Make sure you are building the right it before you build it right. Retrieved from pretotype.org
- Schön, D .A. (1983). *The reflective practitioner*. Basic Books.
- Schön, D. A. (1992). Designing as reflective conversation with the materials of a design situation. *Knowledge-based systems, 5*(1), 3-14.
- se R., Tscheligi, M., & Giller, V. (2003, April). Paper prototyping-what is it good for? A comparison of paper-and computer-based low-fidelity prototyping. In *CHI'03 extended abstracts on Human factors in computing systems* (pp. 778-779).
- Sennett, R. (2008). *The craftsman*. Yale University Press.
- Simon, H. A. (1996). *The sciences of the artificial*. MIT Press.
- Star S. L., Griesemer J. R. (1989). Institutional ecology, 'translations' and boundary objects: Amateurs and professionals in Berkeley's Museum of Vertebrate Zoology, 1907–39. *Social Studies of Science, 19*, 387–420.
- Stompff, G. (2022). Design thinking for professionals: Applied design research as a driving force for innovating education. In Joore, P., Stompff, G., & van den Eijnde, J. (Eds.), *Applied Design Research: A Mosaic of 22 Examples, Experiences and Interpretations Focussing on Bridging the Gap Between Practice and Academics*. CRC Press.
- Stompff, G., & Smulders, F. (2015). The right fidelity: Representations that speed up innovation processes. *Design Management Journal, 10*(1), 14-26.

- Stompff, G., van Bruinessen, T., & Smulders, F. (2022a). The generative dance of design inquiry: Exploring Dewey's pragmatism for design research. *Design Studies, 83*, 101136.
- Stompff, G., Joosten, M., Prince, A., Claessens, M., Geurts, W., & Köppchen, A. (2022b). Touch ground: Introducing design inquiry in higher education. In Lockton, D., Lenzi, S., Hekkert, P., Oak, A., Sádaba, J., Lloyd, P. (Eds.), *DRS2022: Bilbao*.
- Snyder, C. (2003). *Paper prototyping: The fast and easy way to design and refine user interfaces*. Morgan Kaufmann.
- Van Turnhout, K. Jacobs, M. Losse, M. Geest, T. & Bakker, R. (2019). *A Practical Take on Theory in HCI (A White Paper)*.
- Van Turnhout, K. & Smits, A. (2022). *Radio Dabanga: Applied design research in Human Experience & Media Design*. In Joore, P., Stompff, G., & van den Eijnde, J. (Eds.), *Applied Design Research: A Mosaic of 22 Examples, Experiences and Interpretations Focussing on Bridging the Gap Between Practice and Academics*. CRC Press.
- Visser, W. (2006). *The cognitive artifacts of designing*. CRC Press.
- Visser, W. (2009). Design: one, but in different forms. *Design studies, 30*(3), 187-223.
- Wensveen, S., & Matthews, B. (2014). Prototypes and prototyping in design research. In Rodgers, P.A. and Yee, J. (Ed.), *Routledge Companion to Design Research* (pp. 262-276). London: Routledge.
- Wrigley, C., & Mosely, G. (2022). *Design Thinking Pedagogy: Facilitating Innovation and Impact in Tertiary Education*. Taylor & Francis.
- Zimmerman, J., Stolterman, E., & Forlizzi, J. (2010). An analysis and critique of Research through Design: towards a formalization of a research approach. In *proceedings of the 8th ACM conference* (pp. 310-319).

Catelijne van Middelkoop,
Ryan Pescatore Frisk

12. Concerning Apples & Oysters

DOI: 10.1201/9781003491484

Introduction

By far, the best place to train people for practice is in practice (Lawson & Dorst, 2009). The depth and nuance of knowledge imparted by practice-based learning are significant. However, even under ideal conditions, aspects central to the professional design field cannot be replicated in formalised design education. This chapter embarks on an in-depth examination of design education paradigms in the Netherlands, scrutinising the nuanced interplay between theory and practice. The discussion begins within the historical backdrop of art and crafts education, tracing its origins from the guild system's decline to the contemporary challenges faced by creative vocational schools, art academies, and universities. A thorough exploration of educational differences, contextual separations, and the historical evolution of design disciplines sets the stage. The focus narrows onto *Transition Atelier The Last Makers*, a Living Lab embodying a real-world environment for interdisciplinary collaboration and innovative experimentation. As the chapter proceeds, it seeks to unravel the complex relationship between educational levels, bridge gaps in knowledge transfer, and challenge existing paradigms in design education.

Educational Differences

In the Netherlands, design programs are offered at specialised vocational schools (*vakscholen*) and Regional Education Centers (ROCs) in secondary vocational education (mbo), at universities of applied science and art academies (*kunstacademies*) in higher vocational education (hbo), and in scientific education (wo) at general and technical universities. Once a student obtains a secondary school diploma, those who wish to continue their education in the design field have many options to consider. However, not all design programs are equally accessible. Differences between institutions and developmental paths can limit individual potential and hinder social and cultural progress on a larger scale.

One primary influence is reflected in processes of rationalisation, which was increasingly present in the programs of technical schools (Prak, 1979), restricting the learning trajectory exclusively to students possessing a secondary education with an emphasis on science subjects (*exacte vakken*) and fundamental truths, which other institutional levels do not expressly teach. MaxWeber connects rationalisation to bureaucracy, exemplified by the production of social stratification as impersonal, rational processes of social order (Ritzer, 2007). 'Weber is clear that what distinguishes the educated class is its very education and its desire to seek power through this' (Rao & Singh, 2018). 'In line with rationality, bureaucratization was developed to proficiently manage organisations—however, bureaucratization's segmentation and de-personalisation of the work process negatively impact worker's autonomy and creativity. Credentialism, which helps advance bureaucratization, created a system of social closure where only individuals with specific education credentials obtain positions in organisations' (Wui & Leviste, 2023). From this perspective, it is easy to understand how educational stratification reinforces a disconnection and hierarchy of knowledge types.

As a result, people have often regarded design and research as separate endeavours – the former residing in industrial practice and craft, the latter in academic experiments and reflection (Stappers & Giaccardi, 2014). Despite their differences, most settings that offer design programs still share one common characteristic: the physical spaces dedicated to making, which we will refer to as *workshops* for now. The idea of the workshop is used to transcend normative feature comparisons and classifications and instead explore unseen connections, relations, conflicts, and diverse knowledges.

Comparing Fruits

How can we compare the apples and oranges in front of us? Delving into the intriguing mind of taxonomist John Ray, one cannot help pondering his motivation when, in 1670, he added, 'As like an apple to an oyster' to the section Proverbial Similes in his *Handbook of Proverbs* (Ray, 1670). The evolution of this metaphor of dissimilarity over time and in different contexts, transforming oysters into oranges and giving rise to entirely new and unexpected comparisons like 'grandmothers and toads', adds a layer of curiosity to Ray's original intent.

While categorising these specific biological organisms may prompt contemplation, hierarchy is not the sole organising principle at play. At least not in design. The practice in taxonomy, the science of classifying and naming organisms based on shared characteristics. Just like the use of idioms, expressions with figurative meanings that cannot be understood by interpreting their individual words; canons in art history, sets of authoritative works, texts, and principles that are widely accepted as genuine and fundamental; tautologies in math, statements that are always true, regardless of the truth values of their individual components; typologies in things, systematic classifications or categorizations of entities based on shared characteristics; and principles in design, fundamental truths, guidelines, and rules that serve as a foundation for beliefs, actions, or reasoning, all play a pivotal role in creating and fostering understanding amongst groups of otherwise divergent individuals with a variety of skills and knowledges.

Common Roots

In the Netherlands, where making has not yet moved to the core of education (Lehmann, 2020), the roots of art and crafts education trace back to the decline of the guild system, a pre-industrial network of associations involving skilled artisans, craftspeople, and merchants (Prak, 1979). Most contemporary design disciplines originated from this craft-based tradition of creating tangible products (Lawson & Dorst, 2009).

Initially, guilds and crafts were in opposition, with guilds organising independent professionals and crafts forming distinct groups. Craft was the common term for the local trade organisation of individuals engaged in the same (economic) activity, such as weaving (weavers) and tanning (tanners). Drapers and cloth merchants, on the other hand, were members of guilds. Cloth merchants, for example, were traders who sold the cloth but did not manufacture it themselves (Haemers, 2016). Until their abolition in 1798 (Simon Thomas, 2008), the guilds aimed to protect members' interests, uphold craftsmanship standards, and regulate competition. Becoming a skilled craftsperson, such as a cobbler or weaver, typically involved a 3- to 5-year apprenticeship under a master, emphasising practical, on-the-job learning by doing (Lawson & Dorst, 2009).

Over time, guilds evolved from regulating professional conditions to establishing monopolies, influencing social and economic structures. Despite exclusivity, applied skill outcomes had no inherent social difference (Prak, 1979). Stonemasons, not architects, historically led building construction, and blacksmiths and carpenters crafted products, not industrial designers (Lawson & Dorst, 2009).

Crafting furniture, chalices, altarpieces, or statues was considered manual labour, holding equal standing or, later, equal disregard. These activities were categorised as mechanical arts (*artes mechanicae*), learned in workshops and serving practical needs. Which stood in contrast to the liberal arts like geometry and astronomy, taught at universities, primarily fulfilling intellectual needs, and more akin to what is now referred to as science or knowledge (Prak, 1979).

Contextual Separation

During the Renaissance, the work of the mind continued to enjoy a far greater prestige than work done by hand, at least in most disciplines. For painting, sculpture, and architecture, this status changed due to exceptional achievements by artists such as Da Vinci, Donatello, Michelangelo, and Raphael, as well as architect Brunelleschi, who invented linear perspective in the early 1400s and contributed to the evolution of the science of seeing, influencing the dominance of the visual arts. Acknowledging painting, sculpture, and architecture as liberal arts marked the onset of the dichotomy between fine and applied art. Scholars introduced the term *academy* in connection with Plato's school for higher learning. Despite fine artists' daily activities only loosely aligning with Plato's academic vision, they embraced the term, continuing to apply manual skills in established workshops (Prak, 1979), emphasising the concept of *thinking with their hands*.

This contextual separation persisted with the establishment of the Accademia delle Arti del Disegno in 1563 by Giorgio Vasari, which provided a theoretical supplement to the regular workshop training of emerging artists (Pevsner, 1940). Drawing was fundamental to all fine arts, with dedicated space for theoretical and practical drawing classes. While theoretical boundaries were transcended on a conceptual level, practical workshops of applied artists outside the academy maintained a disengagement between different disciplines and between thinking and doing.

Co-creating Quality

In the entrepreneurial environment surrounding guilds and academies, merchants gained control over raw material supply, allowing a single tradesperson to hire multiple weavers who were put to work in independent workshops and deviating from guild wage standards. Conversely, royal businesses established by the French court granted privileges to hired workers, exempting them from guild obligations, with products avoiding quotas. Staff at these royal manufacturers were trained internally, sometimes with contributions from foreign workers bringing innovative techniques. Commissioned artists collaborated on fabric, tapestry, and furniture design alongside their independent work (Prak, 1979).

This artistic practice continues in the Netherlands today, exemplified by Royal (*Koninklijke*) Tichelaar in Makkum and the Vlisco Group in Helmond. Royal Tichelaar, initially a stone factory, now actively collaborates with artists in residence, external designers, and architects. They leverage their extensive knowledge in ceramics and glazes to co-create innovative products, including ceramic skins on building facades (De Vries, 2010).

A similar co-creation process unfolds at the Vlisco Group in Helmond, which has been known for high-quality printed fabrics since 1846. The invention of roller printing made the process less labour-intensive and more efficient than traditional manual methods. Vlisco emphasises the design process, granting commissioned artists ample time for intricate patterns. For instance, artist and designer Michiel Schuurman spent over a year developing a pattern inspired by light-emitting neons together with Vlisco's technicians. According to the website It's Nice That, the work is 'A huge feat considering the design totally disrupts traditional light-on-dark printing conventions' (It's Nice That, 2016). Both contexts highlight the importance of advanced internal communication in bridging the gap between design and execution, ensuring consistent production quality.

Upscaling Taste

A larger scale and more advanced division of labour, with designers working alongside producers, such as the highly skilled technicians still working at Tichelaar and Vlisco today, provided independent companies a significant economic advantage over typical small workshops of guild members. This systemic approach to and partial centralization of production gave the French an artistic advantage over other countries, which they maintained from the 17th to the mid-19th century and showcased in an increasing number of intricately illustrated books. These books allowed craftspeople and ordinary people elsewhere to follow and even acquire French taste. This ambition fed into the desire of well-to-do Dutch who, at that time, preferred to obtain artistic, well-made consumer goods from abroad that were generally considered more appealing than those of Dutch manufacturers (Simon Thomas, 2008). As society continued to industrialise and prosperity increased, a desire to participate in the 'culture of their time' also grew among the middle class elsewhere in Europe. With a booming demand for decorative arts products, national governments and city administrations throughout the continent wanted to limit imports and promote exports. One way to achieve this was through the quality of design (Prak, 1979).

In many places, this resulted in a desire to elevate the level of local manufacturing through better training of both personnel and entrepreneurs. This subsequent focus on educating craftspeople was part of a growing demand for knowledge. Sometimes, separate schools were established for this purpose. Similar to the Accademia delle Artidel Disegno, the education provided by these arts and crafts programs supplemented workshop training and included drawing from examples, plaster, and live models, and often some theoretical instruction in linear drawing and perspective. Practical training was not seen as necessary, as it was already provided in workshops and at manufacturers, reinforcing the gap between design and execution (Prak, 1979).

Sweeping social changes occurred due to the Industrial Revolution, even though industrialisation and modernisation took place more slowly in the Netherlands than elsewhere in Europe (Simon Thomas, 2008). From mechanical inventions to innovative approaches to traditional practices, the most unique was the merger of technology with industry. No doubt impressed by the

French Industrial Exposition in 1844, Henry Cole, one of the editors of the *Journal of Design and Manufactures*, which encouraged artists to apply their designs to everyday articles that could then be mass-produced and sold to the great 'unwashed', convinced the British Prince Albert, husband of Queen Elizabeth, to organise a *Great Exhibition* (Prak, 1979). Where the journal aimed to improve the standards of British industry and provided the middle-class audience with instructions on taste (Coleman, 2001), the exhibition offered an unprecedented chance for all nations to display their best work and compare it with the production quality of other countries.

Humbling Experience

The result of this immense bazaar was, as far as applied or industrial art goes, depressing everywhere (Pevsner, 1940). In particular, the 'tasteful appearance' of what was presented 'left much to be desired' (Van Voorst tot Voorst, 1980). the Netherlands was not at all prepared for such competition. However, it was not the stagnant industry but rather the lack of interest by the Dutch government that was the chief reason for the sparse representation. Compared to other countries, the Netherlands did not consider an excellent international display of its national industry to become a government matter (Simon Thomas, 2008).

Embarrassed (and aware of a missed economic opportunity), the Department of Education, Arts, and Sciences strongly advocated for the establishment of Dutch art academies and schools of applied arts (*kunstnijverheidsscholen*) to improve the quality of applied and industrial arts in the Netherlands. However, because the results of teaching tasteful appearance were not as easily measurable as those coming out of trade schools or technical schools and precise positioning of applied art products in a broader societal context was lacking (compared to, for example, graphic arts in Switzerland or furniture, glass, and ceramics in Sweden), the schools were heavily constrained in their resources during the pre-war crisis years (Prak, 1979). The modest financial resources of the applied arts schools limited their technical capabilities and, consequently, their curriculum. For example, glass was decorated but not blown because the schools needed an oven and blower.

Still, there was a demand for experts. Once again, companies took the initiative to train skilled workers themselves. For example, the glass factory in Leerdam established its own glass school in 1940, and it worked with teachers who occasionally also had teaching positions at applied art schools (Prak, 1979). Sometimes, this resulted in cross-pollination between what happened within workshops at both locations.

The introduction of small melting furnaces democratised glass design and manufacturing, enabling artists to work independently in their own studios. As a glass designer and teacher, Sybren Valkema (1916-1991) recognized this potential at the first American Craft Council (ACC) congress in New York. Following the closure of the Leerdam Glass School, Valkema aimed to integrate studio glass into the curriculum of the IvKNO in Amsterdam. In 1969, he founded the Glass Working Group at the Gerrit Rietveld Academie, emphasising artists' control over the entire glass-making process. Valkema's efforts ensured that studio glass became integral to the academy's curriculum and gained broader acceptance in the Netherlands (Meihuizen, 2009).

Future Practitioners

Another example of an educational program rooted in the outside world can be found at SintLucas in Boxtel, a secondary vocational school that originated from professional demand for future practitioners to strengthen a specialist workforce. Here, members of the union of Dutch painting patrons were searching for the 'Catholic painter patrons of the near future' who would be educated at the first Dutch Catholic School for Painters: St. Lucas. Seen as one of three pillars by the inspector of industrial education (*nijverheidsonderwijs*), ir. G. Slot, who performed the official opening in 1948, this school for painters (*schildersschool*) distinguished itself from the National Painting School in Utrecht by being the first to renew the training system and to convert the winter school into a day school (Nieuwsblad van het Zuiden, 1960). Prior to the foundation of the National Painting School in 1922, there was no specialised training for decorative painters in the Netherlands. Technical vocational education took place in trade schools. At the same time, private, more specialised institutions were established in certain locations to meet the demand for specialists in, for example, faux wood and marble painting (Simon Thomas, 1998).

The program's initial focus on applying paint as protection and finishing was tied to the housing industry and, over time, evolved into a specialised curriculum for aspiring restoration and decoration painters. A growing interest in creativity and aesthetics since the late 1960s influenced a demand for additional courses in commercial art such as advertising (*reclame*), showcase design (*etaleren*), and presentation techniques (SintLucas, 2023). At this time, the curricula of most art schools in the Netherlands moved away from traditional arts and crafts education (Van den Eijnde, 2015) to make room for other approaches. Today, the original restoration and decoration painting program at SintLucas coexists with other programs designed in response to new tools, emerging technologies, and ongoing developments within and demands from professional practice.

This integration of traditional (arts and) crafts and emerging technologies can also be seen at two other creative secondary vocational schools: Cibap in Zwolle and HMC (Wood and Furniture College) in Amsterdam. In response to signals that many studio occupations (*atelierberoepen*) were in danger of disappearing (Consortium Creatief Vakman, 2017), all three schools offer programs in creative craftsmanship (*creatief vakman*) where students are trained in a chosen specialisation such as wood, textile, leather, ceramics, and glass. Although students learn in traditional workshop settings, the program focuses on developing contemporary craftsmanship by incorporating innovative techniques and creative design principles into the curriculum. Students are challenged to work with materials and techniques that are new to them (Consortium Creatief Vakman, 2017), and which could help to position themselves as perspective designer-makers (*zelfproducerende ontwerpers*) in broader and interconnected contexts, on one's own initiative (Huygen, 1984) yet in co-creation with others.

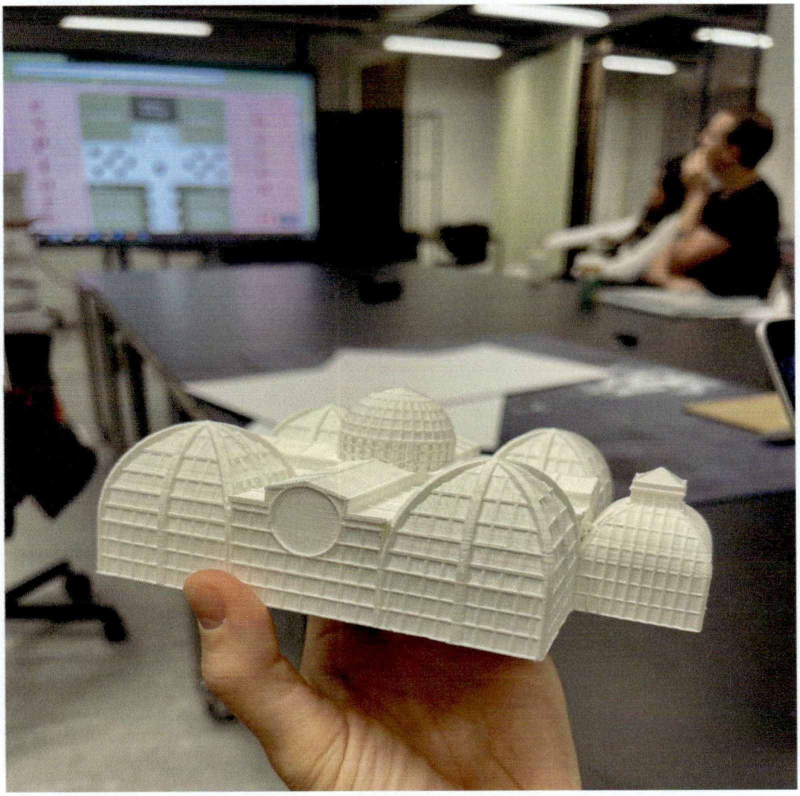

Figure 1. Green Oasis, the rough 3D printed model used for discussion, represents an externalisation of cognition to facilitate knowledge sharing and development in co-creation. (Produced by students in the Context and Situatedness program at the Willem de Kooning Academy in Rotterdam, 2023.)

Contemporary Resonance

Educational workshops, experimental learning environments, and innovation hubs can play a significant role in formal education, professional development, community engagement, and skill-building initiatives. The skills taught in these workshops depend on the context and the reason for their teaching. For instance, there are only a few places in formal Dutch continued education where specific crafts, such as stained glass or restoration painting, are taught. Thus, the sounds of a glass cutter scoring a pattern, sandpaper smoothening a surface, or a palette knife working pigments and medium into a smooth mix might be rare; however, the whirring sound of motors spinning in 3D printers can commonly be heard in workshops at all three levels of Dutch design education.

Although rapid prototyping techniques, such as laser cutting and 3D printing, have replaced some manual labour and brought traditional factory production lines into educational workshops and even student dorm rooms, the output of most consumer-grade fabrication tools, such as 3D printers, lacks the smooth finishing possible with industrial machines. The diversity in materials and possible sizes are limited compared to those available through industrial processes. Still, printed objects can be used in various ways, such as a replacement for a broken part in a coffee maker, a rapid prototype for testing a concept, or an intentional trigger to elicit a response from an audience (Figure 1). While intellectual stimulation from fine art objects may not be associated with immediate practicality, the first two examples can be classified as utilitarian objects that can function regardless of their outward appearance and despite the duration of their intended application.

A lack of attention to the *formgiving* of the object; the process of giving form to that which has not yet taken shape (Ingels, 2020), stands in stark contrast to the care and attention to detail that a craftsperson would take over how something is made and put together.

The skills needed for such a process are developed by actively working and interacting with materials. These materials can range from leather, textile, wood, or glass to 'spider silk' grown in a lab and pictures taken with a digital camera or generated by AI (Wijntjes & Van Middelkoop, 2023). From inception, designing and making can be seen as the translation of ideas into diverse material manifestations in both the physical and digital world.

However, some contend that the concept of craft is limited to application within the tangible realm, exclusively involving the creation of physical objects. This asserts that true craft requires manual production, predominantly by an individual's hands, only occasionally assisted by power tools such as an electric pottery wheel that might be used by a ceramist or a pneumatic chisel employed by a stone worker (Crombez, 2019). This limited view of what craft materials are and which essential techniques apply in the craft-making process feeds into a persistent image of craft as an old-fashioned tradition that does not belong to modern society or relate to any of the 21st-century skills such as creativity and innovation. Instead, 'crafts are said to represent mainly past and repetition of motor skills, not the creation of unique creative ideas' (Seitamaa-Hakkarainen, 2017).

Cognitive Apprenticeship

The act of making is a form of cognitive activity. Research in the humanities and social sciences regarding the generation of knowledge through practical experiences highlights that the process of making can (1) cultivate intelligence, creativity, and skills in individuals; (2) foster social unity across diverse backgrounds; and (3) promote sustainable perspectives on both a local and global level (Lehmann, 2020). In secondary vocational education (mbo), gaining practical experience is an integral part of training. By the time students graduate from a creative vocational school, they will have gained no less than 1,000 hours of direct experience during internships alone (HKU, 2021). This number of hours increases if the hours spent on making in regular classes are included. What is done in all these hours and what is learned depends significantly on the individual student, the task(s) at hand, the guidance involved, and the learning context. Variation in these primary aspects has a significant influence. For example, in Scandinavian countries such as Finland, craft education is an independent and compulsory subject included in the National Core Curriculum for Basic Education (Seitamaa-Hakkarainen, 2017). Here, learning by making, in accordance with Papert's constructionism, posits that learners are not only building abstract knowledge but also creating new artefacts and cultivating new ways of thinking and acting (Seitamaa-Hakkarainen, 2017).

The practice of materialisation is a tool of meaning in the everyday practices of making sense and influencing new meaning in our shared and interconnected semiotic landscapes. 'By engaging in physical activities and producing material products that have meaning for us, we make it possible for those meanings to mediate our future actions' (Lemke, 2000). We can also take advantage of another area of potential that is unique to the context of material production (including visual or digital) in both formalised education and practice by utilising the processes or expertise to externalise cognition for reflection and development (Pescatore Frisk & Van Middelkoop, 2023). Cognitive Apprenticeship (CA) extends learning models from traditional apprenticeship (Collins et al., 1987) as a means to make cognitive and metacognitive processes more visible (Collins et al., 1991). Collins et al. (1987) point to four primary distinctions between traditional apprenticeship and CA: (1) the externalisation of cognitive skills and processes that usually are internalised as an available resource for reflection and development, (2) with a 'relatively transparent' or logical relationship to the outcome, (3) the optimised design and sequencing of tasks, problems, skills, and knowledge in contrast to 'job demands', and (4) the decontextualization or abstraction of knowledge through a diversity of contexts.

In addition to the four points above, situated learning and intrinsic motivation (sub-categories of the sociological principle for designing cognitive apprenticeship environments) are relevant to this discussion. We can explore how contemporary workshops are sites for the production of knowledge. The learning environments utilise situated practice, where new skills, ideas, and approaches are explored, understood, generalised, and abstracted in the contexts of real-life practice and/or using knowledge from real-life experiences. Second, as the materialisation of cognition, iterative cycles explore varied techniques and their implications, building an understanding of meaning potential and the distributed networks that comprise it. A significant tool for transcending pragmatic boundaries is reflected in the third distinction between traditional apprenticeship and CA: 'the optimised design and sequencing of tasks, problems, skills, and knowledge in contrast to job demands' (Collins et al., 1991). The externalisation of cognitive processes in situated practice, and the transfer or generalisation of skills across diverse contexts while resisting (or being able to resist) outcome-centric processes, rationalisation, and normalisation of 'job demands' is a proven catalyst for development and innovation.

Additionally, the foregrounding of production and externalisation reveals and refines attention to discourse (material, digital, or visual). Often, the lifted restriction to 'job demands' allows a sudden broadening of perspective and human connection as the expertise in material production also exerts control on shared networks of semiotic potential. In other words, the relational, material, visual, and digital ideologies that structure our social life and identities enter the experimental workshop and, ultimately, the maker's expertise.

Challenging Constraints

In *Return to Default* and *A Vorkurs for Machines*, the materialisation, tools, and practices are structures that enter into a conceptual, critical, and experiential dialog with contemporary culture. 'While digital infrastructure is often out of our immediate observation, its influence penetrates all facets of life, especially visual and material ideology' (Pescatore Frisk & Van Middelkoop, 2023). *Return to Default* (Figure 2) challenges our expectations of material constraints, simulating the characteristics and operations of digital infrastructure. In *A Vorkurs for Machines* (Figure 5), painting robot BobRob produces material objects that are overwhelmingly non-human; however, because digital infrastructures are intertwined with everyday life, we understand this material representation as a type of stylistic representation or modality tied to technology instead of a generative process devoid of human presence. Suppose BobRob's portfolio was submitted to the Bauhaus as an admission requirement for entry into its foundational course 100 years ago. In that case, we can be nearly certain that this anomaly would be apparent and alien. 'Differences in human culture are regulated, constructed, and mutually influenced by the material discourses that comprise life as we know it. As core aspects of human culture, such artefacts are the media of social structure' (Pescatore Frisk & Pauwels, 2019). 'Global infrastructures of research, design and manufacturing mean that the material world is already understood as a designed world, in other words as a world that mediates between everyday and professional domains, and defines and positions people in political hierarchies and structures, through the qualities of thought and consideration that it embodies' (Drazin, 2020).

Figure 2. 'Just as it is important to recycle materials, Moreno Schweikle and Janne Schimmel believe that it is vital to recycle shapes. They took three functionally perfected yet uninspiring office chair designs, digitally manipulated their DNA, and relaunched them into the physical world' (DAE, 2018).

Return to Default was conceived as a graduation project at Design Academy Eindhoven and built on earlier experiments that took place in the setting of a course that asked students to explore the relationship between humans and machines and the respective role of designers. The students were challenged to explore the possibilities of materialising a digital render in the real world (Figure 3) to evoke or surpass the same appeal as the digital original. This meant that the makers had to convince production companies outside of the institution of the value of their work.

The workshop inside of the school needed to be expanded. Since the exhibition of the results of this challenge during the Dutch Design Week in 2018, the makers still receive occasional requests from private collectors to reproduce their artefact, which is partially done in their own workshop (Figure 4).

Figure 3. 'Three standard office seats have been digitally stretched, blown up and morphed. The off-size frames were 3D-printed, then traditionally upholstered and reintroduced into the office' (DAE, 2018).

The collaboration with the robotics experts at the Applied Labs in Delft was especially valuable during the attempt to create a Bauhaus-inspired *Vorkurs for Machines*. A simple remark from a maker's perspective, such as 'let's have BobRob use oil paint instead, as it dries much slower than acrylic paint,' was seen by an engineer as a Eureka moment. Entirely focused on the hand movement of the painting robot instead of what the hand was doing, the aspect of paint drying too quickly was solved (by the engineer) by shortening the brush stroke and allowing the robot hand to dip into the paint more often. With the paint drying less quickly, the strokes became smoother, but also less authentic.

Figure 4. 'A lean conference chair is blown up into an overstuffed version' (DAE, 2018).

After the experiments, BobRob was moved to an external workshop at Royal Delft (*De Koninklijke Porceleyne Fles*), a producer of high-quality Delft Blue since 1653, that established an experimental department where young potters could experiment with the factory's materials and ovens in 1956. Here, BobRob and five bachelor students from the TU Delft were invited by Head of Design and TU Delft IDE (Industrial Design Engineering) alumnus Joffrey Walonker to take their project *Digital Delft Blue* to a higher level and test the robot's ability to match the skill level of an in-house master painter. Although BobRob, in practice, was far from achieving the level of an expert painter, the robot did reach a point where a practical application for Royal Delft became clear (Mols, 2021), providing underdrawings for human painters—the role of an apprentice.

Experimental projects like this are extremely valuable and equally as rare. Limited access to high-tech tools, such as a robot arm, results in students working in general workshops in design schools not experiencing and perhaps even stretching the limitations of what is possible. External workshops (embedded in companies) not only have more advanced machines and are better equipped, but they also have technological expertise and experience that can come in handy when their processes are reverse-engineered, allowing for the un-making over waste materials, dissecting overproduction in useful components (e.g., materials) that can be repurposed in various ways.

Testing the boundaries of what is possible in manufacturing led to the neck watch of Bruno Ninaber van Eyben, who went public as the first self-producing designer/designer-maker in 1976. One's own needs and/or dissatisfaction with the existing supply have always been a logical reason for designers to produce themselves. These initiatives often lead to series production for third parties. Because a manufacturer takes up the product or due to the circle of enthusiasts expanding, as happened with the fluorescent lamp by Bruno Ninaber van Eyben (Huygen, 1984) and the overstuffed armchair from *Return to Default*.

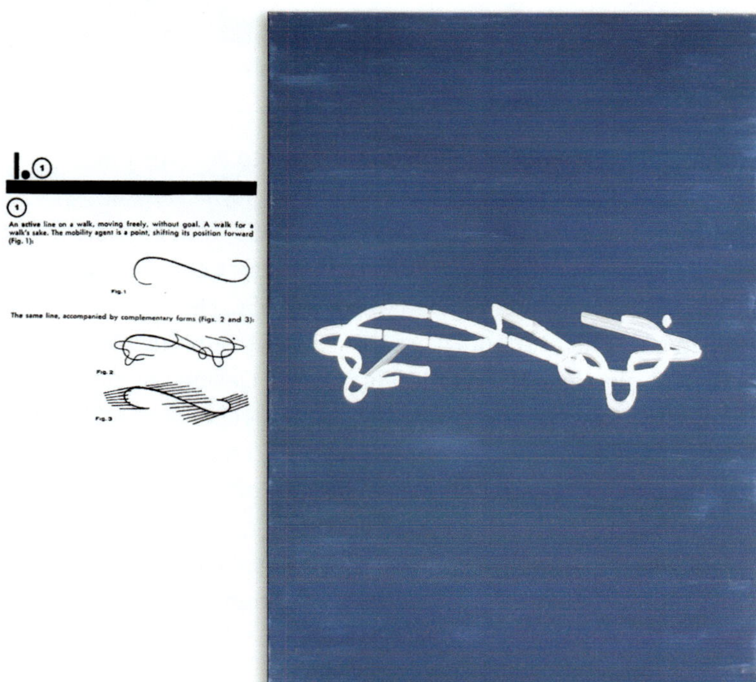

Figure 5. A walk for a walk's sake (2019). Based on a Bauhaus exercise by Paul Klee. The white stroke painted by BobRob is interrupted multiple times in a visible pattern. To avoid the paint from drying too quickly, the robot arm was programmed to get fresh acrylic paint multiple times along the way (Van Middelkoop, 2023).

Facilitating Transitions

Based on the concept of a Living Lab, *Transition Atelier The Last Makers* is a real-world environment, community, and workplace where humans (e.g., students, researchers, instructors, and business employees) collaborate with non-humans to learn from one another, co-create, test, and critically evaluate new and old technologies, material properties, and possible futures in a real-life setting. It serves as a platform for reflection, innovation, experimentation, and learning, allowing for the direct involvement of end-users (lifelong learners) in the design, development, and making process. Like a Living Lab, the transition atelier facilitates the gathering of authentic feedback, insights, and data, which can inform the refinement and improvement of innovations and test-run possible alternatives before wider deployment in education.

The transition atelier fosters interdisciplinary collaboration because, despite their coexistence and cross-pollination of learning in and between programs, their incorporated disciplines and available workshops do not always occur within and in between the workshops mentioned before and not within and among the independent creative secondary vocational schools (mbo) themselves (Van Middelkoop et al., 2024). This is even though an educational system's most essential function must be to facilitate the transition from one layer of expertise to the next (Lawson & Dorst, 2009), which is a challenging task.

For example, even though secondary vocational education (mbo) graduates are expected to be able to move on to higher vocational education (hbo) fluently, many mbo graduates discontinue their studies before they obtain an hbo diploma (Ministerie van OCW, 2023), or do not even consider continuing their education at this advanced level on several grounds; e.g., because a suitable and equally specialised follow-up program does not exist (Van Middelkoop et al., 2024), or because the level is preconceived as too high, time-consuming, or expensive. On top of this, the labour market is favourable (Van den Broek et al., 2020) and, therefore, an attractive alternative for continued education in practice.

Still, a more natural continuation should be possible of learning experiences within creative secondary vocational schools and art academies with applied art and design programs. In addition to their historical roots in arts and crafts education, both art schools and creative vocational schools value self-expression through creative practices. They share the use of studios, workshops, and stations—physical spaces dedicated to making—that are central to creative education and are key places for experimentation. However, relevant innovation resulting from actual (applied) design research within these experimental learning and innovation environments is often overlooked (Van Middelkoop, 2022), undervalued or disregarded, and it is sometimes seen as original rather than functional (Ribbens, 2012). This is despite practice-based research being a primary source for development and innovation, also if it starts with simple curiosity about the properties of a material or a subjective personal experience.

A complicating factor is that research is only officially conducted on two of the three educational levels currently in the Netherlands, within higher vocational education (universities of applied science; hbo), and scientific education (universities; wo). Legally, secondary vocational education (mbo) has no role in research. At the same time, mbo closely aligns with the business sector, where significant innovation occurs. Because of this, outgoing Minister of Education, Culture, and Science Robbert Dijkgraaf, in 2023, announced his plans to give secondary vocational education (mbo) a full role in innovation and research (ScienceGuide, 2023).

This does not mean that research has not already been taking place within this setting. Following the successful development of practice-based research professorships or lectorates in hbo and the strengthening of their positioning in the official research landscape that used to be dominated by universities, 88 so-called practorates have been set up during the past ten years, and twelve are currently in the process of formation (Practoraten.nl, 2023). All practorates have in common that they aim to bridge the gap between education, research, (regional) business, and the future creative workforce. At creative secondary vocational school, SintLucas' Practorate, Meaningful Creativity (*betekenisvolle creativiteit*) is done through Research through Design, which in this context takes a prominent role in establishing ways to prepare for the future through (practising) traditional crafts. In order to do so, the practorate is expanding the practice of doing research as a (recognized) part of the process of designing products (and services) into the design of activities and artefacts that serve as crucial components in the process of generating and communicating knowledge (Stappers & Giaccardi, 2014).

Crossing Boundaries

With its practorate, SintLucas has established its own experimental learning and innovation environment in previously unexplored territory. Fostering an unconventional approach to long-term problem-solving, this creative space encourages individuals to make bold decisions, even in the face of uncertainty. In pursuit of this objective, the protectorate is currently preparing the next step: a pilot project named Transition Atelier The Last Makers (*Transitieatelier De Laatste Makers*), in which creative thinking will be extended into critical doing and a future-proof curriculum centred around (un)making will be co-created.

In order to facilitate this, the practorate has extended the idea of boundary objects (including project briefings for students) by addressing the potential of boundary materials to generate new ideas, locate unseen knowledges, and create alternative routes to the future (Van Middelkoop & Van Harn, 2023). Materials, in all [their] resilience and autonomy (Focillon, 1934), and not limited to the ones that are the central focus within the creative craftsmanship programs offered at SintLucas in Boxtel, play a vital role in this dynamic process. A process that is not primarily focused on solving short-term problems like 'how to get rid of abundantly and commercially produced residual materials' but rather on reconsidering possible futures based on sustainable decisions made during the (design) process of making.

In order to do so, new perspectives on collaboration, in which the material is an active participant, are urgently needed. Like many other collaborations between educationalists and researchers, the outcome and follow-up depend on social resources such as time, space, and the quality of interactions among all parties involved. Although the ultimate intention remains to bridge the gap between education, research, (regional) business, and the future creative workforce, the learning process so far has primarily taken place within the context of SintLucas itself, not beyond.

Shaping sustainable collaborations requires attention and patience. In practice, gradually seeking justification for choices and critically reflecting on them often proves challenging. However, when designing collaborations that contribute to knowledge development and provide space for education, it is crucial to continue to think further about their application and long-term consequences.

Exploring the material in alternative ways (Figure 6)—from using placebos, e.g., scrap leather as a temporary stand-in for future alternatives such as cell-cultured collagen-based leather (Gerritsen, 2018), and digital skins in 3D modelling software, where the material can be questioned and applied in exaggerated quantities and without hindrance—gives rise to new roles while established ones fade away, making the meaning of creativity and critical craftsmanship in sustainable collaborations increasingly clear.

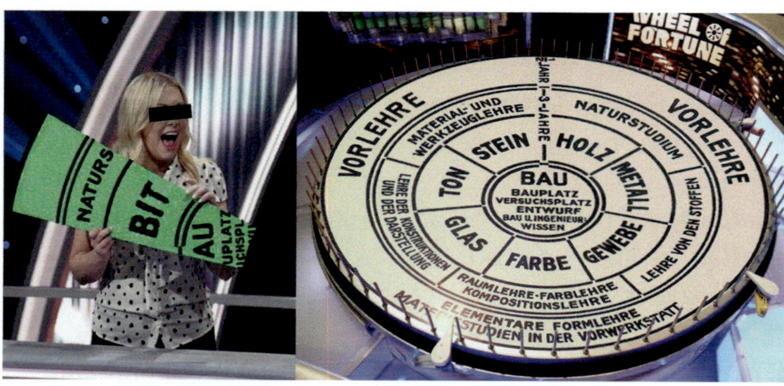

Figure 6. Data as material. Modification of the diagrammatic representation of the Bauhaus syllabus by Walter Gropius (Van Middelkoop, 2019).

Bibliography

- Broek, A. van den, Cuppen, J., Ramakers, C., Termorshuizen, T., & Vroegh, T. (2020). *Dalende doorstroom mbo-hbo: waarom stroomt een steeds kleiner aandeel van de mbo-studenten door naar het hbo?* Nijmegen: ResearchNed.
- Brown, J.S., Collins, A., & Duguid, P. (1989). Situated Cognition and the Culture of Learning. *Educational Researcher, 18*(1), 32-42.
- Clarke, M., & Clarke, D. S. (2010). *The Concise Oxford Dictionary of Art Terms*. Oxford University Press. https://doi.org/10.1093/acref/9780199569922.001.0001

- Collins, A., Brown, J. S., & Holum, A. (1991). Cognitive apprenticeship: Making thinking visible. *American Educator, 15*(3), 6-11, 38-46.
- Collins, A., Brown, J. S., & Newman, S. E. (1987). Cognitive apprenticeship: Teaching the craft of reading, writing and mathematics (Technical Report No. 403). BBN Laboratories, Cambridge, MA. Centre for the Study of Reading, University of Illinois.
- Consortium Creatief Vakman. (2017). *Creatief Vakman*. https://creatiefvakman.nl/index.php/consortium-creatief-vakman-2/
- DAE, Man and Communication. (2018). *Return to Default*. Dae.wiki. https://www.designacademy.nl/p/study-at-dae/graduation-show/graduation-projects/janne-schimmel-return-to-default
- De Vries, M. (2010). *Represent Koninklijke Tichelaar*.
- Drazin, A. (2020). Design Anthropology in Context: An Introduction to Design Materiality and Collaborative Thinking (1st ed.). Routledge. https://doi.org/10.4324/9781315688732
- Focillon, H. (1934). *Vie des formes*. Paris: Librairie Ernest Leroux, Presses Universitaires de France.
- Gerritsen, W. (2018, June 15). *Maastrichts biotechbedrijf kan leer kweken*. De Limburger. https://www.limburger.nl/cnt/dmf20180614_00064062.
- Haemers, J. (2016). Ambachtslieden in de stad (uit: 'Gouden tijden', ed. V. Lambert & P. Stabel). Kuleuven. https://www.academia.edu/26722152/Ambachtslieden_in_de_stad_uit_Gouden_tijden_ed_V_Lambert_and_P_Stabel_
- HKU. (2021). *Interview met Marieke Gervers*. https://www.hku.nl/het-werk-van-hku/verhalen/interview-met-marieke-gervers
- Huygen, F. (1984). Tussen ambacht en industrie 1. *Items 14*. https://itemsmagazine.com/1984/10/1/tussen-ambacht-en-industrie-1/.
- It's Nice That. (2016, May 12). *Michiel Schuurman's giant illustrations meditate on the tiniest of details.* https://www.itsnicethat.com/articles/michiel-schuurman-illustration-vlisco-120516
- Ingels, B. (2020). *BIG. Formgiving. an Architectural Future History*.
- Lawson, B., & Dorst, K. (2009). *Design expertise*.

- Lehmann, A.S. (2020, March 1). Curious hands: moving making to the core of education. *NWO*. https://www.nwo.nl/en/projects/cisckc205
- Meihuizen, J. (2009). *Glas(s), expositie en boek over 40 jaar glas aan de Rietveld Academie*. Designgeschiedenis. https://www.designhistory.nl/2009/glass-expositie-en-boek-over-40-jaar-glas-aan-de-rietveld-academie/
- Ministerie van Onderwijs, Cultuur en Wetenschap. (2023, April 19). Doorstroom studenten mbo. *Middelbaar Beroepsonderwijs | OCW in Cijfers*. https://www.ocwincijfers.nl/sectoren/middelbaar-beroepsonderwijs/studenten/doorstroom-studenten-mbo#:~:text=Rond%2030%25%20van%20de%20instroom,haalt%2C%20rond%20de%2050%25
- Mols, B. (2021). *Paint robot Bob Rob serves Delft Blue apprenticeship.* TU Delft Stories. https://www.tudelft.nl/en/stories/articles/paint-robot-bob-rob-serves-delft-blue-apprenticeship
- Nieuwsblad van het Zuiden. (1960, September 14). Schilderspatroons der naaste toekomst in nieuw gebouw.
- Pescatore Frisk, R. & Pauwels, L. (2019). The shared stories that written words tell when no one is reading: exploring modality in typographic landscapes as an ecosocial semiotic system. *Visual Communication*. https://doi.org/10.1177/1470357219860086
- Pescatore Frisk, R. & Van Middelkoop, C. (2023). Situating the (Un) Common Landscapes of Shared Futures: Developing Speculation as a Co-Design Framework of Cognitive Apprenticeship to Empower Diverse Stakeholders and Contest Bias. *Research and Advances in Education, 2*(12), 41–65.
- Pevsner, N. (1940). *Academies of art, past and present*.
- Practoraten.nl (2023). *Practoraten op de kaart*. https://www.practoraten.nl/practoraten/
- Prak, N. L. (1979). *Geschiedenis van het ontwerp-onderwijs*. Cantecleer bv.
- Rao, S. & Singh, S. (2018). Max Weber's Contribution to the Sociology of Education: A Critical Appreciation. *Contemporary Education Dialogue, 15*, 73-92. https://doi.org/10.1177/0973184917744964
- Ray, J. A. (1670). *Handbook of Proverbs*.

- Restauratoren Nederland. (2023, September 18). *Opleiding en bijscholing – Restauratoren Nederland.* https://www.restauratoren.nl/over-restauratie/opleiding-en-bijscholing/
- Ritzer, G. (2007). The Weberian Theory of Rationalization and the McDonaldization of Contemporary Society.
- Ribbens, A. (2012, October 13). *Design dat er toe doet.* NRC. https://www.nrc.nl/nieuws/2012/10/13/design-dat-er-toe-doet-12563466-a30610.
- Rutten, P.W.M, & Cramer F., Chabot J., & Troxler P. (2013). Reinventing the art school, 21st century. Rotterdam University of Applied Sciences. https://www.rotterdamuas.com/research/projects-and-publications/pub/reinventing-the-art-school-21st-century/71fc5168-4e25-4b18-b7c7-efba2042ea5f/
- Simon Thomas, M. (1998). De Leer van het ornament: Versieren volgens voorschrift, 1850-1930, *Studies in the Decorative Arts, 6*(1), 127–131. https://doi.org/10.1086/studdecoarts.6.1.40662670
- Simon Thomas, M. (2008). *Dutch Design. A History*.
- SintLucas. (2023, November 17). *SintLucas 75 jaar*. https://www.sintlucas.nl/sintlucas-75-jaar.
- Stappers, P. & Giaccardi, E. (2014, January 1). *Research through Design*. Interaction Design Foundation – *IxDF*. https://www.interaction-design.org/literature/book/the-encyclopedia-of-human-computer-interaction-2nd-ed/research-through-design.
- Van den Eijnde, J.N.M. (2015). *Het huis van ik : ideologie en theorie in het Nederlandse vormgevingsonderwijs*. ArtEZZ Press. https://scholarlypublications.universiteitleiden.nl/access/item:2890489/view
- ScienceGuide (2023, September 19). *Dijkgraaf wil zijn waaier doortrekken naar het onderzoek in het hbo en mbo*. https://www.scienceguide.nl/2023/09/dijkgraaf-wil-zijn-waaier-doortrekken-naar-het-onderzoek-in-het-hbo-en-mbo/.
- Van Middelkoop, C. (2019). 'Bauhaus', *'Anderen bekeken ook' – Het belang van het Bauhaus voor het ontwerponderwijs van morgen.* In Symposium nederland bauhaus Museum Boijmans van Beuningen.

- Van Middelkoop, C. (2022). Un-/Blackboxing the creative potential of Dutch secondary vocational education. In *Collaboration for Impact* (pp. 72-73). https://www.nadr.nl/wp-content/uploads/2023/02/NADR_CollaborationForImpact_RGB.pdf
- Van Middelkoop, C. (2023). De Laatste Makers. *SintLucas.* https://online.fliphtml5.com/acdtk/jukw/#p=1
- Van Middelkoop, C., & Van Harn, R. (2023). Leren van Zink? De rol van grensmateriaal in tegenspel. In M. Buizer, J. Schueler, & L. van Arkel (Eds.), *Transitiezones: navigeren op weerstand*. *Accelerating the Circular Economy Zuid-Holland (ACCEZ).*
- Van Middelkoop, C., Van Harn, R., Roebroek, D., Koenen, L., & Jonkers, S. (2024). *Onze Taal*. SintLucas.
- Van Voorst tot Voorst, J.M.W. (1980). Nederland op de Wereldtentoonstelling van 1851 te Londen. *Nederlands Kunsthistorisch Jaarboek, 31*, pp. 475-492.
- Wijntjes, M., & van Middelkoop, C. (2024). A framework connecting vision and depiction. *Visual Cognition.* Teaching Sensation and perception. https://doi.org/10.31234/osf.io/geyht
- Wui, M.G.L., & Leviste, E.N.P. (2023). Weber, Education, and the Rise of Modern Societies. In: Geier, B.A. (eds) *The Palgrave Handbook of Educational Thinkers*. Palgrave Macmillan, Cham. https://doi.org/10.1007/978-3-030-81037-5_116-1

In Conclusion

Peter Joore, Anja Overdiek,
Wina Smeenk, Koen van Turnhout

13. Unlocking the Potential of Living Labs: Insights and Strategies

Epilogue

DOI: 10.1201/9781003491484

Introduction

What are the lessons that applied design researchers can take away from this book? What suggestions and recommendations can be derived from the experiences presented in the preceding chapters? On the one hand, it is clear that working with Living Labs offers unique opportunities to address some of society's ambitious challenges. At the same time, it is also clear that it's not necessarily that straightforward to deal with the dynamics of a Living Lab.
In Chapter 1, *Experimentation at the Heart of Societal Change,* we presented our expectation that applied design researchers can play an important role in connecting different levels of the Living Lab and its surroundings. To explore this, we first discussed the Living Lab and its relationship with societal change processes. Second, we discussed the social dynamics that take place in the Living Lab itself. Third, we discussed the actual practices of applied design researchers within a Living Lab, for instance with regard to their engagement with tangible forms. In this final chapter, we present some of the overarching insights that can be drawn from the many reflections in this publication.

Enabling a Customisable Approach

From the preceding chapters, it has become clear that Living Labs as experimental settings can definitely contribute to the realisation of societal transitions. Just as there is a wide variety of societal challenges, there is also a wide variety of associated Living Labs. After all, transitions in the energy sector require very different contextual considerations than those in elderly care, or those in agriculture. Overdiek and Van der Laan therefore argue in Chapter 2, *Living Labs and Other Experimental Environments,* that it is necessary to recognise this diversity. While being transparent about applying basic elements of the Living Lab like co-creation and real-life settings, it is essential for applied design researchers to adapt their approach and methods to the specific context of the experiment. Every Living Lab is unique, and applied design researchers need to adapt their repertoire accordingly. Unsurprisingly so, we also find a wide variety of approaches in this book. For example, in Chapter 9, *Bridging Multi-Stakeholder Dialogue about AI Systems in the Lab,* Van der Horst et al. present Living Labs as a real-world context to discuss emerging technologies with diverse stakeholders. We also recognise this argument in Chapter 10, *Between Experiments – Leveraging Prototypes to Trigger, Articulate, and Share Informal Knowledge,* in which Jaśkiewicz and Smit present the case of the Cities of Things Living Lab. Here they consider Living Labs primarily as a context for research through design. Other authors, besides emphasising the future-oriented aspect of Living Labs, specifically unpack their potential to deal with the complexity of society. For instance in Chapter 5, *Exploring the Potential of Festivals as Living Labs for Systemic Innovation,* Dijkstra et al. suggest that labs can model the systemic nature of societal challenges, and that festivals can provide a unique testing ground for innovations aimed at initiating systemic change. It is clear that there are many types and forms of Living Labs, linked to different experimentation methods and goals, and related to different types of societal challenges. Thus it is essential for design researchers to adapt their methods and activities to these specific settings.

Embracing Real-World Complexity

It is almost a truism to state that Living Labs relate to the real world, but the chapters in this book indicate that this relationship is multifaceted and sometimes challenging. Van den Eijnde and Mohammadi, for example, argue in Chapter 3 that labs should embrace real-world complexities. This call resonates in many forms throughout the other chapters in this publication. However, incorporating or excluding complexities is riddled with complex decisions. This becomes clear from their chapter, in which they highlight the challenges that occur when setting up Living Labs that are really significant and meaningful. Embracing complexity means that applied design researchers need to consider which are the elements they want and can incorporate at a specific stage, and what complexities are better left out for the moment. This is also discussed in Chapter 6, *Opening & Closing Hours,* in which Smeenk et al. emphasise the importance of making conscious choices about the moment to open up a lab to the real world, and when to shield it from the outside environment. This can be done multiple times, depending on how the co-design process evolves and which unexpected events and insights emerge. They indicate that it is essential to consciously think about which elements are open to the outside world, and which aspects are not. This does not only apply to outside contextual complexities, but also to intra-personal ones. In Chapter 4, *Co-designing towards Transitions?,* Sluijs et al. stress the importance of reflecting different levels of awareness that stakeholders bring into the lab, and to align them with the co-design methods used in the lab. That these choices are non-trivial is also voiced in Chapter 2, in which it is explained that applied design researchers may risk overlooking the disparities between concrete innovation projects and the societal transitions that they intend to accelerate. In general, we might say that every lab setup exhibits a profound effort to 'be real' in some sense and needs to compromise real-world complexity in others.

Balancing Risk and Safety

Whereas it is non-trivial to embrace the complexities of the real world within a Living Lab, it can also be necessary to keep the real world outside the door, at least temporarily. After all, for experimentation to be successful, it is essential that a certain degree of safety is present. Of course, this precondition applies to conducting material experiments, as it would be unfortunate if injuries occur because a new technology turns out to be dangerous for the user. But this need for safety applies at least as much to social aspects and dynamics. Can employees or citizens freely express their opinion, even if this opinion goes against the ideas of a higher ranking manager or the government official they are depending on? And vice versa, can eldermen or managers make themselves vulnerable, without the risk of people taking them less seriously afterwards? To innovate, it is essential to facilitate taking risks, but these must be affordable risks. Living Labs are the perfect place to encourage this kind of calculated risk-taking. The setting of the lab, the internal culture, the way of working, and the practical facilitation can all help to create the safe space needed to come up with new and unexpected alternatives to an existing situation. This is for instance addressed in Chapter 5 in which the authors argue that the temporary and physical boundaries of a festival's environment can provide unique opportunities for exploring radical innovations. In Chapter 2 the authors explain that user-centred approaches and co-creation strategies are essential to engage stakeholders in the kind of thinking and doing necessary to initiate effective change. The challenge then is to facilitate stakeholders in adopting a creative and open 'designer mindset' that embraces uncertainty and complexity. It is important to be aware of the efforts and risks that less powerful stakeholders who are often citizens might face, versus the efforts of professional actors facing less personal risk. In summary, Living Labs are about encouraging risk-taking in a relatively safe and ethical way.

Encouraging Creative Collaboration

Virtually all authors indicate that engaging relevant stakeholders, and facilitating collaboration between them, is an essential component of Living Labs. For example, the authors of Chapter 5 emphasise that collaboration between different types of stakeholders can foster innovation and experimentation at festivals. Similarly, in Chapter 11, *Ceci n'est pas un Prototype,* Stompff et al. suggest that the disciplinary diversity of stakeholders in a Living Lab is an important ingredient for improving its innovation potential. As do the authors of Chapter 9 when they explain that understanding systemic aspects of emerging technologies such as their ecological footprint, and creatively engaging the stakeholders in it, is crucial for effective co-creation in Living Labs. But how to deal with the fact that many partners are not trained as designers at all? Designers can foster creativity by challenging existing perceptions within the context of the Living Lab. In Chapter 3 the authors explain that designers should prioritise co-creation with end-users and stakeholders, in order to foster a sense of shared ownership among stakeholders. Similarly in Chapter 6 the authors suggest that stakeholders should have the opportunity to be involved in collaborative decision-making in the context of Living Labs. It is up to the facilitator or team to streamline and coordinate the dynamics associated with such shared decision-making processes. The designer's unique ability to approach an issue from different perspectives can be invaluable here. If the designer then also manages to encourage other stakeholders to view the collective challenge from a different perspective, this will further increase the chances of productive collaboration.

Facilitating Rich Learning

Productive collaborations are not possible without making sure that labs form an environment in which everyone can learn. This means, first, acknowledging the type of knowledge and experience all stakeholders bring into the lab. In Chapter 10 the authors argue that both formal and informal knowledge coexist in the lab. The learning perspective is further articulated by Van Turnhout and Andriessen in Chapter 7, *Experimenting with Novel Knowledge.* In their chapter, the authors discuss different ways in which individual learning and collective learning can be aligned. They suggest that innovative environments where individual learning is combined with collective learning are a promising approach to the deployment of Living Labs. The value of learning also emerges in Chapter 6 where the authors analyse three real-life cases based on the different learning mechanisms arising from boundary-crossing theory. They suggest that joint learning processes are a key enabler of the success of Living Labs. This is in line with Troxler and Mostert-van der Sar, who emphasise the importance of serendipitous learning in Chapter 8, *The Open Lab as Boundary Object*. They suggest that transdisciplinary encounters are key to spark innovations if stakeholders are willing to learn from each other. Also the authors of Chapter 4 argue that designers need to develop new skills in order to facilitate this rich learning. In conclusion, it is clear that an open attitude and approach of both individuals and collectives facilitate joint and rich learning. Such an attitude and approach are essential to make sure the cultural and epistemic diversity of a Living Lab environment can effectively be brought to fruition.

Creating Tangible Artefacts

The development of material objects and settings is a key ingredient of imaginative collaborations. At the same time, creating tangible and visible innovations is one of the designer's significant distinguishing qualities. Designers can create a new reality which did not exist before, through the synthesis of seemingly disconnected elements towards a new whole. In Chapter 12, *Concerning Apples and Oysters,* Van Middelkoop and Pescatore Frisk elaborate on the development of the environment in which creative professions, also called studio or workshop professions, take place. From this perspective, Living Labs are actually variants of the long-established workshop or atelier of the painter, sculptor, architect, and other creative professionals. When it comes to Living Labs, the challenge for the applied design researcher is to collaboratively organise this creation process, involving all other stakeholders. The joint development of material objects and settings, designed boundary objects, provotypes, prototypes, technological trials, and ideal types can help to find and explore collaborative reflections. It can encourage new ways of thinking about the situation at hand. For example in Chapter 8, the authors unpack how practical experience and experimentation can promote understanding and innovation. In Chapter 11 the authors suggest that tangible objects actually represent future realities or ideals. They emphasise that these objects can significantly promote effective discussions among stakeholders. Tangible objects can also play a productive role in the social dynamics of Living Labs. For instance, in Chapter 9 it is being described how the use of tangible objects can make difficult topics negotiable, and in Chapter 10 the authors suggest that prototypes can be a catalyst for collaboration and experimentation within Living Labs. Making sure that these objects are accessible and meaningful to all parties involved will promote transparency and alignment. It is clear that the act of materialising settings and idea directions, in various shapes and forms, is an important part of the applied design researcher's toolbox for working with Living Labs.

Future Considerations

While the aforementioned lessons underscore the considerable potential of Living Labs, they also bring to light the challenges and risks that demand attention. Important issues are related to the long-term impact, the scaling, the ethics and values, the power dynamics, the facilitator roles, and the reflexivity and reflection in Living Labs. Each of these challenges may point to key elements of the future research agenda for applied design researchers that are active in this field of expertise.

An important consideration is related to the *long-term impact* of Living Labs. While it's essential for Living Labs to continually assess initial tangible results and insights, in line with their learning-focused nature, equal emphasis should be placed on evaluating the lasting impact of collaborative endeavours. This involves prioritising transformational learning and sustainable change over the long haul, potentially necessitating the tracking of outcomes and the adaptation of strategies to ensure ongoing success. For this, looking deeper into collective Living Lab programs where different kinds of labs would be, simultaneously or successively, working together on a transition challenge will be necessary.

The long-term impact is closely related to the importance of the *scaling* of Living Lab learnings, products, and processes. Whereas market-focussed perspectives on scaling rely on *'scaling up'* and *'scaling out'* at the end of a project, it is important for applied design research to explore other forms of scaling along the process of experimentation. For instance, the concept of *'scaling in'* may offer a promising notion related to the dissemination of transformational concepts and worldviews in an organisation or broader societal system. One question could be how such a perspective could take place, and could even be accelerated, in a Living Lab project. Also the concept of *scaling scree* may be a relevant one, where a commitment to small incremental innovations adds up to a bigger shift towards a systemic change. Applied design researchers could find out more about these alternative ways of scaling and play with them.

Another future consideration is related to the issue of *ethics and values*. When engaging stakeholders from various sectors of the quadruple helix (business, knowledge institutions, government, and citizens), careful consideration must be given to their roles and capacities to contribute, take risks, and wield influence. It is obvious that all involved parties should be treated with respect, striving for a balanced approach towards both contributions and benefits of each party. While financial support may be a necessary precondition for many applied design researchers to be involved in a Living Lab project at all, this may also create an unconscious bias towards themes and subjects for which sufficient funding is available. As external funding is often supplied by some of the most powerful stakeholders, for instance government or large organisations; this may steer researchers in their decision of what subjects to tackle, and what subjects to ignore. When looking at ethics and values in a broader, more-than-human sense, applied design researchers need to scrutinise how to engage all relevant 'voices' and stakes in a lab, also those of biological non-humans and ecosystems.

The previous subject is closely related to the subject of *power asymmetries*. As mentioned already, different actors bring in different contributions, for instance finances, time, or specific insights. Applied design researchers should actively engage with the power imbalances that may occur within a Living Lab. While many labs emphasise the importance of co-creation, it is not self-evident that transparency in the decision-making process has been properly addressed. This need by no means be a conscious omission, as it is often implicitly assumed that collective enthusiasm is enough of a prerequisite for good decision-making. By ensuring that all participants have equal opportunities to contribute, are included in the decision-making process, and have access to relevant information, applied design researchers can effectively deal with the unconscious bias that may arise from the difference in power between various actors.

When regarding the role of the applied design researcher as a *facilitator*, it's vital to strike a balance between presence and absence. While facilitating active participation, they should avoid becoming overly present or indispensable in the process. When being engaged in the development of a Living Lab, applied design researchers may consider how they could for instance gradually render themselves redundant over time, as other stakeholders become more and more self-sufficient in their collaborative efforts. A distinction between more strategic roles in and across Living Labs, and the intricate task of facilitating valuable moments of collaboration and creation, is helpful to fine-tune and 'hand-over' roles between applied design researchers and other partners in a multi-stakeholder group.

Another future consideration of applied design researchers has to do with the need for *reflexivity and reflective practices* within participatory design processes. This entails taking time for critical reflection on the learning that occurs and the dynamics among different stakeholders. By promoting reflexivity, one can ensure that learning is actively cultivated rather than taken for granted. Encouraging reflective practices among stakeholders, both individually and collectively, may involve dedicating time for introspection and dialogue, as well as providing support for the formation of empathy and understanding of diverse perspectives.

With this chapter, this book on Living Labs and other Experimental Learning and Innovation Environments has come to an end. In other words, it is time for the authors, and the readers, to enjoy a moment of reflective practice themselves. In these last lines of this final chapter, we can look back on an inspiring collective learning process. And, if we have then enjoyed this moment of reflection long enough, perhaps we can already start thinking about the next knowledge cycle of the Network Applied Design Research, which will focus on ways to further increase the societal impact of applied design research. But let's not get ahead of ourselves. For now, we trust that lessons drawn from this publication can make a valuable contribution for anyone engaged in Living Labs, and we hope that after reading this book, methods and activities related to experimental environments will be even more meaningful in the future!

About the Editors

DOI: 10.1201/9781003491484

Peter Joore

Peter Joore is professor of Open Innovation at NHL Stenden University of Applied Sciences, Leeuwarden, the Netherlands. He is chairing the Network Applied Design Research. His research focuses on the relationship between new incremental innovations and long-term societal transition processes. Peter has an Industrial Design Engineering background and started his career in industry, working on innovations ranging from the design of aircraft interiors to the development of airport signage. He worked as senior researcher and business consultant at TNO, the Netherlands Organization for Applied Scientific Research, where he worked on sustainable system innovation, mass customised solutions, and wearable sensor systems. In parallel to his work at TNO, he supported the start-up of the new faculty of Industrial Design of the Eindhoven University of Technology. Peter holds a PhD from the faculty of Industrial Design Engineering of Delft University of Technology, where he is still a visiting researcher at the Design for Sustainability group.

Anja Overdiek

Anja Overdiek is professor of Cybersocial Design at Rotterdam University of Applied Sciences and associate professor at the Mission Zero Center of Expertise at The Hague University of Applied Sciences (the Netherlands). Her research focuses on multi-stakeholder co-design towards systemic change, on experimental environments, and on digital design to support sustainability transitions. She co-authored two practice-oriented books about Living Labs and is a founding member of the Expert Network for Systemic Co-design (ESC). Anja holds a PhD from Freie Universität Berlin (Germany) in Political Sciences and approaches her research with a systemic and 'more-than-human' design lens.

Wina Smeenk

Since 2021, Wina Smeenk has been a professor in Societal Impact Design at Inholland University of Applied Sciences. She is also chair of the Expertnetwork Systemic Co-design (ESC), a network of four universities of applied sciences in the Netherlands. Next, Wina launched her own empathic co-design agency 'Wiens Ontwerperschap' in 2010. She graduated from Delft University of Technology, where she studied Industrial Design, after which she spent over 25 years working as a strategist, co-designer, and researcher for a variety of international businesses, government, and non-profit organisations. Moreover, she helped to develop innovative design-oriented educational programmes. These include universities of applied sciences, THNK, the Amsterdam School of Creative Leadership, and the Faculty of Industrial Design at Eindhoven University of Technology. In 2019, she defended her PhD thesis 'Navigating empathy, empathic formation in co-design processes'. Wina authored the books, *The Co-Design Canvas* and *Design, Play, Change, a Playful introduction to Design Thinking.*

Koen van Turnhout

Koen van Turnhout is professor of Human Experience & Media Design at Utrecht University of Applied Sciences. His research group examines the design of digital media. In particular, how smart, data-driven solutions can be designed in a human-centred way and how people can interact with those devices while maintaining human autonomy. He has a background in human-computer interaction, multi-modal interaction, design, and research methodology. Koen obtained his PhD in Industrial Design at Eindhoven University of Technology.

About the Authors

DOI: 10.1201/9781003491484

Marije Boonstra

Marije Boonstra has a multidisciplinary background in psychology, culture, and media studies. She is currently working as a researcher and student coach at NHL Stenden University of Applied Sciences. Her interests range from education, creativity, and culture to sustainability. Together with Aranka Dijkstra, she was author of the Festival Experimentation Guide, a practical guide for sustainable innovators on designing, implementing, and evaluating experiments at festivals.

Aranka Dijkstra

Aranka Dijkstra looks to accelerate sustainable transitions through implementing real-life experiments and interdisciplinary collaborations. Since 2014, she has been guiding students and entrepreneurs in designing and implementing experiments in various Urban and Festival Living Labs. Even after many years, Aranka maintains a strong connection to festival experimentation and aims to spread her knowledge and experience with festival experimentation through various publications and in particular the Festival Experimentation Guide.

Morgan Duta

Morgan Duta is designer, researcher, and lecturer at the Designing Value Networks professorship, part of The Hague University of Applied Sciences' Knowledge Centre Mission Zero. She explores what skills designers need to be great facilitators and investigates how design education can support that process. Morgan has a background in design thinking, design research, experience design, service design, product design, social innovation, and education.

Jeroen van den Eijnde

Jeroen van den Eijnde was trained as product designer at the Arnhem Institute of the Arts and as design historian at Leiden University. He holds a PhD based on his research about ideology and theory in Dutch design education. Since 2016, he has been professor of design at ArtEZ University of the Arts, involved in research and innovation projects on sustainable design for textiles, products, and (interior) architecture. He leads the research community NewTexEco for developing new ecosystems for a sustainable textile sector. He's program member of CLICKNL, the Dutch knowledge and innovation network for the top sector creative industries.

Maaike Harbers

Maaike Harbers is a professor of Applied Sciences in Artificial Intelligence & Society at Research Center Creating 010 at Rotterdam University of Applied Sciences. Her work focuses on the ethical and societal implications of artificial intelligence, and she researches how designers of AI applications can account for the implications of their concepts on human values like privacy, autonomy, and inclusivity. Maaike holds a PhD in Artificial Intelligence and a master's degree in Philosophy.

Donagh Horgan

Donagh Horgan is an academic and practitioner in the area of regenerative placemaking and socio-spatial transformation. At Inholland University of Applied Sciences, he is a lab lead at the Urban Leisure and Tourism lab in Rotterdam Zuid looking at societal transitions with a quintuple helix of local partners. Trained as an architect and service designer, he is an expert on social innovation and resilience in the built environment working with communities in the European neighbourhood and around the world. Alongside academic work, he consults with local government on European projects as an expert on programmes such as URBACT, and he supports cities with urban innovation through collaboration with the Centre for Public Impact on projects for Bloomberg Philanthropies and UNDP.

Tiwánee van der Horst

Tiwánee van der Horst is a researcher at the Research Centre Creating 010 of Hogeschool Rotterdam for the research groups Artificial Intelligence & Society and Cybersocial Design. Tiwánee has a background in architecture and visual arts. In the digitalising world, she investigates how technology can be designed to benefit both people and planet. She does this by making tangible the invisible social and ecological aspects of technology. For instance, she designed the object 'Who pulls the strings of AI?', which is about the power relations behind AI. This led to research into more-than-human design principles for designing tangible objects used for systemic dialogue.

Mark Jacobs

Mark Jacobs (Inholland University of applied sciences) specialises in crafting meaningful learning experiences within higher education. As a researcher and lecturer in the field of design thinking, he delves into the intersection of creativity, innovation, and education. At the Centre of Expertise for Creative Innovation (CoE CI), Mark contributes to collaborative projects involving diverse stakeholders – academia, industry, government, and the public – as they address intricate challenges within the vibrant Amsterdam Metropolitan Region. Additionally, he also is Project Lead for the Design Thinking Open Resources Platform (DORP), an initiative that curates a collection of open educational materials centred around exemplary prototypes and practical interventions fostering imagination in (higher) education.

Tomasz Jaśkiewicz

Tomasz Jaśkiewicz is professor of Civic Prototyping at the Rotterdam University of Applied Sciences and Design Fellow in Prototyping Complexity at the Faculty of Industrial Design Engineering at the TU Delft. His core belief is that technological innovation should be an inclusive and democratic process. To this end, his research revolves around developing new tools, methods, and strategies supporting designers and non-professionals in iterative prototyping with digital technologies. His work is strongly influenced by his background in architecture and urban design, experience of running and failing a startup, and a lifelong passion for building quick-and-dirty interactive prototypes.

Elise van der Laan

Elise van der Laan is a sociologist and associate lecturer in Fashion & Design Research at ArtEZ University of the Arts. She obtained her PhD in 2015 from the University of Amsterdam for a study into aesthetic standards in fashion photography. She has published on topics such as fashion and identity, aesthetic standards, sustainable fashion, and cultural fields.

Ju Laclau Massaglia

Ju Laclau Massaglia is a Research and Communication intern at the Centre of Expertise Mission Zero and about to graduate with a bachelor's degree in International Communication Management. Within her work with Future-Proof Labs, she collaborated in the writing of the *Innovating with labs 2.0: Creating space for sustainability transitions* book.

Catelijne van Middelkoop

Catelijne van Middelkoop is practor (research professor) at SintLucas, and Context Developer at the Willem de Kooning Academy. In these roles she curates experimental learning environments aimed at fostering knowledge through making. Additionally, Catelijne is a design fellow and co-director of the Pictorial Research Lab at TU Delft's Faculty of Industrial Design Engineering where she previously served as full professor of Visual Communication Design. In her co-directorship of Strange Attractors, a design research studio, Catelijne specialises in developing innovative methods, tools, and critical perspectives to produce insight towards measurable change. Concurrently, she pursues a PhD at the University of Groningen.

Masi Mohammadi

Masi Mohammadi is a smart living scientist with broad experience in people-centric and health-promoting environments. Leading the 'Empathic Environments' research program, she integrates societal changes with AI and robotics into architectural systems, creating an enriching and inclusive built environment that enhances living quality. She serves as a full professor and chair at both Eindhoven University of Technology and HAN University, focusing on next-generation smart homes and neighbourhoods. As the scientific director of the DEEL Academy, she drives real-world, collaborative projects, merging academic research with community needs. Her work, recognized around the world, involves extensive collaboration to develop innovative housing solutions.

Manon Mostert-van der Sar

Manon Mostert-van der Sar is the dynamic director and co-founder of Stadslab Rotterdam, the FabLab of Rotterdam University of Applied Sciences, and is also the author of *Hey Teacher, Find Your Inner Designer.* An advocate for cutting-edge education, she not only trains teachers and coaches instructors but also designs cross-institute modules, all while nurturing her passion for hands-on teaching in the lab. Beyond academia, she's a hands-on maker, rebuilding her farmhouse with bricklaying, sawing, milling, and carpentry – a testament to her practical creativity. Her multifaceted approach to education and DIY ethos embodies a commitment to both professional excellence and personal growth.

Ryan Pescatore Frisk

Ryan Pescatore Frisk leverages decades of renowned critical and experimental practice as an award-winning designer, creative director, developer, producer, and cultural anthropologist to address rapid and emerging changes in the production of meaning and identity in everyday landscapes through biases in sociotechnical systems. At the Visual and Digital Cultures Research Center (ViDi), University of Antwerp, as a research lecturer at Willem de Kooning Academy, and a co-director of Strange Attractors, Ryan's comprehensive foundation of multiple specialisations with industry-leading transdisciplinary practice and research expertise empowers diverse human and non-human knowledge towards methods, tools, and perspectives that contest bias and prefigure futures.

Perica Savanović

Perica Savanović is a professor of Built Environment at the Inholland University of Applied Sciences. He was born in the former Yugoslavia and brought his background in electrical engineering to the Netherlands. Perica studied architecture at the Delft University of Technology and, after graduation, went to work at the architectural firm of Van den Broek en Bakema. Fascinated by the difference in thinking between design and engineering, he obtained his doctorate with a thesis in Integral Design at the Eindhoven University of Technology. To apply what he had researched in practice, he then went to work as program manager for Integral Collaboration at the Construction Research Foundation, and later on as a Development Manager of Integral Housing at the University of Utrecht. In his practice-driven research, started at Avans University of Applied Sciences, Perica explores the relationship between humans and technology and strives to achieve participating creation through communal conceptualization.

Janneke Sluijs

Janneke Sluijs focuses in her research on the role of the facilitating designer to bring together worlds and worldviews of diverse stakeholders in Living Labs. She works at The Hague University of Applied Sciences as researcher with the research group Designing Value Networks and as lecturer at the Master Next Level Engineering.

Iskander Smit

Iskander Smit is the founder and chair of the Cities of Things Foundation, a research program that started at TU Delft, where he was a visiting professor. It now serves as an independent knowledge platform in cooperation with academic partners and industry. Iskander is deeply involved in exploring the relationship between human and machine intelligence from a design perspective. He founded Target_is_New, and he is a curator and board member of ThingsCon. Prior to this, he held positions in strategy and innovation and was involved in various technology initiatives and events as an organiser and public speaker.

Guido Stompff

Guido Stompff is professor of Applied Sciences at Inholland University of Applied Sciences. He is specialised in design thinking/doing and advocates for the relevance of design skills for any professional. His research focuses on finding pathways of action to democratise design. In his design career spanning 25 years, he won numerous design awards. In his academic career, he obtained a PhD on innovation-in-the-wild, with a focus on collective intelligence. He leads a team that focuses on participatory approaches to innovation. He is engaged in networks as Expertise Network Systemic Co-design (ESC) and the Center of Creative Innovation (CoECI).

Sybrith M. Tiekstra

Sybrith M.Tiekstra has a background in arts, culture, and media studies. She has worked on various projects relating to sustainable innovation, entrepreneurship, and the creative sector. Sybrith is interested in the social aspects of sustainable innovation and has a firm belief that creativity and art can play an important role in the sustainable transition. The potential of festivals for innovation is a red thread in her work.

Peter Troxler

Peter Troxler is professor of Revolution in Manufacturing at Rotterdam University of Applied Sciences. He obtained his PhD at ETH Zurich, at the intersection of work psychology and business administration, specialising in organisation design. He has worked as a management consultant at a design consultancy firm in Switzerland (1997-2018), as a research manager in artificial intelligence at the University of Aberdeen, Scotland (2001-2004), and as a senior project manager and freelance executive editor at *Waag* in Amsterdam (2007-2010). He was also the founder, mentor, and an inspiration for over a dozen FabLabs in Europe (2009-2013), including the one in Rotterdam. He worked as producer for an independent theatre company in Switzerland (1994-2001) and was director of a critical artistic research collective in Aberdeen (2003-2007). He currently is the director of the Research Centre Creating 010 at Rotterdam University of Applied Sciences.

Marieke Zielhuis

Marieke Zielhuis works at the Research group Research Competence at the University of Applied Sciences Utrecht. She obtained her Master Industrial Design Engineering at Delft University of Technology. She has worked in various roles which reside between higher education and practice: in Centres of Expertise, in education programs for practitioners, and in research collaborations with practice. Her PhD research focused on the collaboration between research and design practice partners and the contribution of research projects to design practice.

Daan Andriessen

Daan Andriessen is professor of Research Competence at University of Applied Sciences Utrecht. He specialises in optimising the impact of practice-based research and the competence development of researchers at universities of applied sciences. He has a background in administrative science, methodology, and knowledge management and a PhD in Economics from Nyenrode University. His mission is to support researchers in doing research that matters: practice-based research that contributes to the development of knowledge, people, products, and organisations. He has a special interest in the dilemma between rigour and relevance.

Maria Arias

Maria Arias is an Industrial Designer, User Experience Designer, and Design Researcher. Maria's goal of research is to foster positive change through promoting habits and collaborative efforts. Her background includes experience in designing and facilitating co-creation sessions for impact generation and developing design theories and interventions to promote healthy habits. Maria currently teaches and researches UX, co-design, persuasive technologies, and systemic transitions at the Hague University of Applied Sciences.